A FUNNY THING HAPPENED ON THE WAY TO THE GRAVEYARD

ALEX STEWART

© Copyright 2005 Alex Stewart.
All rights reserved. No part of this publication may be reproduced, stored in a retrieval system, or transmitted, in any form or by any means, electronic, mechanical, photocopying, recording, or otherwise, without the written prior permission of the author.

Note for Librarians: A cataloguing record for this book is available from Library and Archives Canada at www.collectionscanada.ca/amicus/index-e.html
ISBN 1-4120-2974-0

Printed in Victoria, BC, Canada. Printed on paper with minimum 30% recycled fibre. Trafford's print shop runs on "green energy" from solar, wind and other environmentally-friendly power sources.

TRAFFORD
PUBLISHING™
Offices in Canada, USA, Ireland and UK

This book was published *on-demand* in cooperation with Trafford Publishing. On-demand publishing is a unique process and service of making a book available for retail sale to the public taking advantage of on-demand manufacturing and Internet marketing. On-demand publishing includes promotions, retail sales, manufacturing, order fulfilment, accounting and collecting royalties on behalf of the author.

Book sales for North America and international:
Trafford Publishing, 6E–2333 Government St.,
Victoria, BC V8T 4P4 CANADA
phone 250 383 6864 (toll-free 1 888 232 4444)
fax 250 383 6804; email to orders@trafford.com

Book sales in Europe:
Trafford Publishing (UK) Limited, 9 Park End Street, 2nd Floor
Oxford, UK OX1 1HH UNITED KINGDOM
phone 44 (0)1865 722 113 (local rate 0845 230 9601)
facsimile 44 (0)1865 722 868; info.uk@trafford.com

Order online at:
www.trafford.com/04-0802

10 9 8 7 6

Acknowledgements

My thanks are due to Mike Blissett for producing the illustrations and the cover pictures, to Anne Bentley for producing the scripts and, particularly, to my wife Janet for putting up with me when my thoughts have obviously been elsewhere.

My thanks are also due to the various people who related many of the amusing incidents to me and whose names I have long since forgotten.

Foreword

"I wouldn't do that," said the Scots minister, gazing out over Sorrento to the blue Mediterranean. (We were having a holiday there.)

I had just suggested that I might like to write a book on some serious topic.

"You would be much better to write down some of those funny stories you've been telling us for the last week or so. Some people might read those."

Some interesting events and funny stories do seem to have come my way during my life, so perhaps he is right. At any rate, I have taken his advice.

Although many of the incidents are true, I have changed or disguised the names of the characters or organisations where embarrassment might arise. To avoid any misunderstanding, there are no people involved whose names are actually Smith or Jones!

Prologue

Davy was finding his way home one night after a heavy session in the pub in Bangor, Co. Down. He lived a mile or so from the town on the road to Newtownards and was using his usual reliable guide and support - the local aristocrat's demesne wall. This extended almost continuously from near the pub to a point where it made an abrupt right-angle turn to the left. Davy's home was just across the road from there so, provided he had hung on tightly to the wall, he was home – safe if not exactly dry. The only break in the wall was presented by the gates in front of the old Abbey Church, but these were usually shut at night. On the night in question, however, someone had left the gates open and, as a consequence, Davy found his way into the churchyard and eventually collapsed in a drunken stupor among the gravestones.

At that time, transport between Bangor and Newtownards was provided by a sidecar - or an Irish Jaunting Car as it is usually called in England. (The piece of coachwork fastened onto the side of a motorcycle still lay far in the future.) This daily service started down at the harbour where the fishermen lived and, on arrival at the Abbey Church, the driver blew a few blasts on a trumpet to attract any possible passengers from the hand-weavers' cottages in the nearby Church Street. On the morning after his drinking session, Davy was wakened by the blast from the side-car driver's trumpet. He sat up, looked at the gravestones around him and announced,

"Well, I'm here, Lord, but I can't remember dying".

I heard this story from my father but it probably stems from my grandfather if not, indeed, from my great-grandfather - an Alexander Stewart like myself and one of the hand-weavers who worked in their own homes in Church Street near the old Abbey Church.

A bit of background

The hand-loom weaving trade must have come to an end in my great-grandfather's time because my grandfather Thomas Stewart did not adopt his father's profession but became a gardener and grave-digger. By his day, Bangor was developing fast. Many fine houses had been built near the sea to house wealthy families with businesses in Belfast and my grandfather was gardener to one of the "nobs". At one point, this nobby gentleman stood as a candidate for parliament and, while canvassing, came into my grandfather's house. Approaching him with outstretched hand, he said,
"Ah, there you are, Tommy".
My grandfather kept his hands behind his back.
"Mr Jones", he said, "You pass me in the garden of your house at least twice every day and never as much as say `You're there, Stewart'. Now you come into my house saying 'Ah, there you are, Tommy'. If it is any satisfaction to you, I shall be voting for you because you are standing for the right party but, for now, I should be just as pleased if you would leave my house."
A prickly lot, the Stewarts. Surprisingly enough, I don't think he got the sack!
My grandfather's wife Mary came from Coleraine, not far from Portrush and the Giant's Causeway. I suspect that the next story originally came from her although I heard it from my father.
One of the first electric tramways in the British Isles was built between Portrush and the

Giant's Causeway. It aroused great suspicion and dislike among some of the locals who, apart from finding the clatter it made rather annoying, suspected that it might be the work of the devil because there were no horses to pull it. Not only that but - horror of horrors - it ran on Sundays for the benefit of the tourists who wished to visit the Giant's Causeway. The line passed a Presbyterian church and the story goes that the Presbyterian minister was indulging in one of his (long) extemporary prayers in which, among other subjects, he dealt with the horrors of this machine, its suspicious motive power and the fact that it ran on the Sabbath.

"Not only that," he said, "it makes a frightful noise."

At that moment, the tram passed by. The minister extended his arms to heaven and exclaimed:

"There you are, Lord, you can hear it for yourself."

It must have been around the turn of the century that my grandfather was approached by another gardener in the town with a business proposition. At that time, it had been realised that tomatoes were edible and not merely ornamental. The other gardener proposed that they should go into partnership, build some greenhouses and produce tomatoes. Tommy Stewart would have nothing to do with such new-fangled and passing fancies so the other chap went ahead on his own. When I was a boy, the other family owned a large area of greenhouses and - at least compared with the Stewarts - were very wealthy.

An incredibly thrifty man, Tommy's sole luxuries in life were his weekly plug of tobacco and going to the football match each Saturday. At some time before the First World War, he built a house on the opposite side of Church Street from the one that he had taken over from his father and moved into the new house with his family. Later, he built a house next door and, when getting on a bit - round about 1925 - he was again in a position to build. At this time, he had four grown-up children, of whom one was a bachelor. He used his money to build two more houses in the garden of the houses he already owned - so that he could leave one house to each child. That would ensure that he would leave no money that could possibly be spent in some irresponsible way!

"The best laid schemes of mice and men ..."

The two new houses were just about finished when the bachelor son died. My grandfather spent the rest of his life worrying about how he was going to divide four houses among three children - and never solved the problem. His will, which had been written before he built the extra two houses, caused some trouble among the surviving children after his death. As a result of this, my father always told his brood that he would do his best to give us a reasonable start in the world educationally but after that, it was up to us. He said that if there were anything left in the kitty when he was getting near his end, he would blue it in.

"Nobody is going to quarrel over anything I leave!"

(He didn't quite manage it but we certainly did not quarrel!)

My family

And so we come to my father, James Stewart. Not, I hasten to add, the film star. James left school at the age of fourteen and became a telegram boy at the post office in Bangor. From there, he progressed to being a postman, then a sorting clerk and, eventually, a counter clerk. At some point, he was moved from Bangor to the head office in Belfast and worked there for some years before being transferred back to Bangor. While he was still working in Belfast, the family moved back to Bangor (to occupy one of the houses built by my grandfather) and James (always known as Jimmy) was a commuter for a few years until - to the detriment of his career but to his great relief - he was transferred back to the post office in Bangor. I don't think he ever liked the post office very much - the bureaucracy did not appeal to his independent mind - but he certainly preferred the small office in Bangor to the large organisation in Belfast. He did, however, have a few amusing tales to tell about his time in the city office, where promotion was largely a matter of waiting for someone to die or retire (whereupon everyone on the list moved up one place).

Two of his colleagues went swimming, in Bangor as it happens, during a day off. One of them got into difficulties but fortunately the other was a trained lifesaver. With some difficulty, he brought the drowning man to the beach and saved his life by artificial respiration. In due course, he was awarded the medal of the Royal Humane Society and some high official of the post office made the presentation

in the presence of a large number of the staff. Calls were made for the hero to make a speech.
"Well," he said, "It's this way. Joe was lucky. He's below me on the list."

It was while he was in Belfast that Jimmy Stewart was promoted to sorting clerk. There was often a flat spot during the evening shift in the sorting office and some of the more adventurous members of the staff would sometimes slip out to the pub across the road for a quick drink. Sam, one of these adventurous spirits, was departing on the illegal errand one evening when another member of the staff, a notorious scrounger, called after him,
"Leave me one at the bar, Sam."
Sam did not deign to reply. On the way over to the pub, however, he had a little think. When he had ordered his drink, he spoke to the barman very quietly and said,
"Listen, there is no need for any panic but I am one of the keepers from Purdysburn."
(Purdysburn was the local mental hospital, still called a lunatic asylum in those days.)
"Now look, one of our patients has escaped. He is quite harmless but he has a fixed illusion that someone has left him a drink at this pub. I'm going off duty now but the chap who is taking over from me should be here any moment. If the patient should arrive in the meantime, just keep him here till the other keeper arrives."
Returning to the sorting office, he told the scrounger that he had left a drink for him at the pub. Later, and diligent, enquiries by other

members of the staff elicited what had happened. The scrounger arrived and said to the barman,

"I believe I've been left a drink here."

He was served immediately by a very attentive barman who, at some detriment to other business, engaged him in conversation. Having finished his drink, the - by now - centre of all attention said,

"I really will have to be moving along. They'll miss me, you know."

"Not at all," said the barman, "Have another drink."

Eventually, despite all the barman's efforts, the "patient" escaped and returned to the sorting office to announce,

"That's a very decent barman in that pub. Do you know, he even bought me a drink."

One rather less amusing thing happened to my father during his period in Belfast. He felt extremely and repeatedly ill and reported to the post office doctor. (In those days, the post office employed its own doctor who had to agree before

anyone was allowed to go sick.) The post office doctor prescribed aspirin and sent my father back to work. As a result, my father's appendix burst and he had to be rushed into hospital with peritonitis. He was very lucky to survive and, if I have got my dates right, I am therefore very lucky to have existed at all! However, even this stay in hospital provided my father with a nice tale. The matron – of the real old-fashioned matron type - would come round every morning to inspect the ward and make sure that everything was spick and span. She would then enquire of each patient in turn - in a very plummy voice,
"How do we find ourselves this morning?"
The patients would all dutifully answer,
"Very well, thank you, Matron."
The man in the next bed to my father was a rather more robust character and, after matron's departure, would rail at the others,
"Very well, indeed! The whole lot of us together wouldn't make a healthy man."
Eventually, one morning, when the matron made her usual pronouncement
"How do we find ourselves this morning?", he retorted,
"Och Matron, I just threw aside the sheet and there I was!"

Another reason why I may be lucky to exist is that when my father, along with many others, volunteered to join the army on the outbreak of the First World War, he was turned down because he was too small. Otherwise, he would probably have been with the Ulster Division, which was more or less decimated at the Battle of the Somme.

Although I was born in Belfast, the family moved back to Bangor while I was still very young and my earliest memories are of Bangor. At the time of the move, I had an elder sister, Marion; a younger brother, Clifford, arrived quite a few years later.

Rather belatedly, let me introduce my mother. Eva Seal was English and during the early part of the First World War, she was in service with a family in Camberley in Hampshire. This family had some Irish connections and decided to move over to Bangor during the war because food was more plentiful in Ireland. Eva agreed to accompany them for six weeks to help them settle in. However, apart from some holidays and a longer stay during the Second World War to help nurse her mother, who was dying of cancer, she never returned to England. She had met my father, who was still working in Bangor and delivered the post to the house where she was working. In later years, she used to relate – with a quiet smile - what the cook in the establishment had said,

"Jimmy Stewart! A fortnight will do you with him. He's the worst flirt in Bangor."

But Jimmy's flirting days were over.

Schooldays

At that time, children in Northern Ireland went to school when they were four years old and, if they only received elementary education, stayed at school until they were fourteen. My younger brother must have had the main outlines of the system explained to him before he went to school because when my father was leaving him in the school playground on his first day at school, the lonely-looking little lad said,
"You will come back and fetch me when I'm fourteen, won't you."
My father nearly took him home again.

There were, if I remember right, about fifty children per class when I was in the elementary school. We learnt our tables by singing them out together (led by the teacher) and we learnt to read in a similar fashion, singing "C-A-T cat" and so on. An absolutely shocking system which would, I suspect, destroy any creative tendencies - according to modern educationalists. Nevertheless, we were reasonably literate and numerate by the age of seven or eight and some of us, at least, were soon driving the staff of the local Carnegie Library round the bend by our continual reappearance at the children's section to change our books. Streaming was simple. The first two classes at the school were called Junior Infants and Senior Infants and that was where the rudiments of arithmetic and reading were learnt. The brighter youngsters, however, went straight from Junior Infants to First Standard and were therefore a year ahead.

My father's earnings from the Post Office were supplemented by my mother taking in "visitors" during the summer. Bangor was very popular with Scots people and people from Lancashire and Yorkshire. Needless to say, the extra work kept my mother very busy and one day she had to send me (despite my lack of years) to the butcher's shop to buy the meat - a rolled roast. On arrival at the shop, I presented the butcher with the half-crown and requested a Rolls-Royce. This may have been an omen, as will appear later.

The "visitors" business must have been a reasonably successful operation because my father was able to buy a second-hand Austin Seven in or around 1936. He paid £45 for the car (which was a 1934 model with very few miles on the clock) and was taught to drive by the garage owner as part of the contract. I always had grave doubts about the effectiveness of the second half of the contract.

The arrival of the "wee" car brought technology into my life. In the first place, I drove a car for the first time when I was about twelve years old and, in the second place, I started to spend quite a bit of time at the garage across the road, where my father bought his petrol and had the car serviced. I was allowed to do little jobs at the garage - such as draining sumps, a task for which my still diminutive bulk must have made me quite suitable. I must have frequently arrived home filthy but my mother was a very tolerant woman! Vic, the owner of the garage, took quite a lot of trouble to explain to me the intricacies of the internal combustion engine so I acquired an early interest in engineering matters.

Vic had a mother-in-law who was a terrible back-seat driver. One day, in the course of his business, he had to deliver a new steering wheel to a customer and was, at the same time, giving his mother-in-law a lift in the back seat. He had the spare steering wheel beside him at the front of the car. Eventually, the lady started her usual practice of back-seat driving so, on a straight piece of road where it was safe to do so, Vic picked up the spare steering wheel and, turning round to the back seat, he handed it to her, saying,

"Here, drive the bloody thing yourself."

The poor woman nearly had a heart attack but the back-seat driving was reputed to have been successfully cured.

For some reason or another, the school chose me as one of the pupils to be given the honour of going to Belfast for the Silver Jubilee visit of King George V and Queen Mary. We were carefully lined up along a road somewhere and given a flag to wave. After standing in the sun for about an hour (luckily it was not raining), we were rewarded by an approximately one second glimpse of a beautiful big car, which I admired very much, and - of course - the King and Queen.

About this time in my life, I joined - or was volunteered for - the choir in the Abbey church. This had two important effects. Although I did not stay in the choir at the Abbey until my voice broke - the choir master at the Abbey church was a rather bad-tempered man and my father agreed to my leaving the choir after some incident or other - I did acquire a love for choral singing which lasted throughout my adult life. The other effect was that I learnt, at an early age, that things are not always what they seem. As we processed down the aisle in our Eton collars, surpluses and cassocks, we would often hear the old dears say things like,

"Aren't they little angels?"

In fact, we were a right bunch of little horrors - which may have had something to do with the choirmaster's bad temper!

There was one lady in the congregation who had a regular place near a window on the right-hand side of the church. Even when the church was packed, it was noticeable that none of the chairs within a radius of about eight feet of her were ever occupied. On judicious enquiry, I was informed that eight feet was considered to be beyond the maximum flea jumping distance. Still, she was a very regular member of the congregation and no doubt the authorities at the Pearly Gates would easily deal with a trivial matter like fleas.

The rector had a very refined accent with the result that my grandmother always referred to him as "the man with the foxy (Oxford) voice".

In due course, I found myself in the "Scholarship Class" at the elementary school. This was the special group of those whom the teachers thought had a chance of winning a scholarship to the grammar school. This class was taken by the headmaster, Mr MacDonald. He was, I am sure, a very able teacher but all I can remember about him is comments by the pupils to the effect that the Campbells should have done a more thorough job. I do, however, remember doing homework till midnight at times to keep up with the demands of the scholarship class (this at the age of ten). Be that as it may, it got me through the scholarship examination.

Until about the age of fourteen, I must have been a somewhat sickly (or perhaps spoilt!) child. At any rate I seem to have seen the doctor quite a lot. There were two doctors in the practice and they had, roughly speaking, a division of labour. One of them dealt very effectively with the more down-to-earth patients (like us) and the other specialised in bedside manner for the wealthier parts of the population. (I am probably being unkind but it improves the story, which I got from my father. How he got hold of it, I cannot imagine.) At any rate, an extremely wealthy and very hypochondriac lady - from a "good" area some distance away on the outskirts of the town - arrived at the practice one day in her chauffeur-driven car. Unfortunately, the bedside manner specialist was not there for some reason and she got the other bloke, who greeted her with,

"Have you come in the car, Mrs ...?"

She agreed that she had.

"Well," said the doctor after a very preliminary examination, "I think perhaps, in the circumstances, you should send the chauffeur back home."

The lady, anticipating a very prolonged consultation, willingly complied. When she came back in, the doctor asked her if she had sent the car back home and, on her reply to the affirmative, he said,

"Good, now if you will just walk back home and then walk about the same distance every day, you will be all right."

Returning to matters scholastic, the scholarship boys joined the grammar school at the age of eleven or twelve. In the three years between then and sitting for the Northern Ireland "Junior Certificate" they learnt practically nothing new except French, Latin, a certain amount of English literature and some science. They had already had to learn most of the maths, grammar, geography, history, etc in order to pass the scholarship examination.

In Form 3A, we were taught maths by Mr Johnston (Johnny), whose real subject was geography. He was no great shakes at geometry but he was a great sportsman. When some geometry proposition had to be proved, he would struggle away for a considerable time till the blackboard was just about completely covered and, with luck, he would get the right result in the end. (The only conceivable reason why he did not simply copy the proof from prepared notes was that he was, indeed, a great sportsman and rose to the challenge.) At this point, one of the clever-clog scholarship boys would pipe up from the back row,
"Please Sir, wouldn't it be quicker to do ...?"
Johnny would then hand over the chalk to the clever clogs, who would rub out a minuscule area of Johnny's working and then do the necessary proof in about three lines.
One thing we were encouraged to do at Bangor Grammar School was to think for ourselves. During prayers on the first day at school after war broke out, Mr Wilkins the headmaster told us that one of the things that we should remember was that "Truth is the first casualty in war." He taught Latin and one of the books we had to deal with was Book Two of Caesar's Gallic Wars – the Vercingetorix one. He explained why the book was written in the third person (as if someone else had written it) and started with great praise for the opposing leader (although Caesar always won in the end!), the reason being that the book was written like this because Caesar was running for election as

Emperor (those may not be his exact words but that was the general drift).

The history teacher told us at an early date that he had not the slightest intention of teaching any Irish history. If he tried to teach us the truth to the best of his ability, he was bound to get into trouble with some or other of the parents. He pointed out that there was usually only one question on Irish history in the Junior Certificate examination and if we wanted to be able to answer that one, the safest thing to do was just read the appropriate chapter in the official text book. (This had the effect – possibly intended – that the more inquisitive among us visited other sources, such as the town library, and learnt all sorts of interesting things about Irish history.) In due course, he departed for the war and after the war he became a Labour MP and eventually a member of the House of Lords.

The English teacher was an Oxford man – a real Oxford man in that apart from having studied at Oxford University, he actually belonged to the town. (I often wonder whether he had to live in college despite the fact that he had a perfectly good bed at home.) However, he did his bit to counteract the bigoted religious background of many of the pupils by making great use of Fitzgerald's translation of the Ruba'iyat of Omar Khayyam. By getting us to analyse the grammar of some of the verses, he ensured that some of the message would stick in our heads. However, I very much doubt whether our school version included the verse which might be regarded as the punch line:

"Oh, Thou, who Man of baser Earth didst make
And who with Eden didst devise the Snake.
For all the Sin with which the Face of Man
Is blackened,
Man's forgiveness give – and take!"

(Similarly, our versions of Shakespeare were interspersed with lines of dots to the exclusion of some of the best lines. On attending a performance of Macbeth after leaving school, it was interesting – and indeed useful - to find that the words replaced by dots in the drunken porter's comments on the evils of alcohol were, in fact:

"And take lechery, alcohol increases the desire but spoils the performance.")

After Junior Certificate, life became a bit more earnest in more ways than one. In the first place, the work was all new from then on and, in the second place, life in general was becoming much more serious. The year was 1940, the phoney war was over and the Battle of Britain had just been won. The names of old boys who had been killed in various fields were mentioned at school prayers more and more frequently as time went on.

Bangor did not suffer much in the way of air-raids during the war. In fact, I think only one bomb was dropped on the town, this occurring during one of the big raids on Belfast. There were two big raids on Belfast both, probably, intended for the shipyard and/or the aircraft factory. The first raid missed its target completely and the bombs landed on some of the poorer areas of Belfast causing a very large number of casualties and much destruction of property. The second raid did, however, do a lot of damage to the shipyard.

When, in the middle of the night, the news of the first raid arrived in Dublin, the official in charge of the fire brigades - without consulting any higher authority - immediately ordered the main fire engines and their crews to head for Belfast and see what could be done to help. More senior officials might have worried that this could breach the neutrality of the Republic of Ireland but the junior official thought human lives were more important.

The arrival of the Dublin fire brigades – which were painted green - caused quite a sensation in Belfast and gave rise to one rather amusing story. Some of the Dublin chaps were engaged on rescue work and were trying to reach a man who was buried under the rubble. Eventually, when they got near enough, the man beneath the rubble shouted,

"Who's there?"
"The Dublin fire brigade."
"God, that was some bomb".

The chemistry practical examination, which I took for the Senior Leaving Certificate, sticks in my memory for two reasons. In the first place, I made a complete hash of a straightforward titration and, in the second place, the inspector from the Ministry of Education, who was there to see fair play, was a man called Foster. Later on, when I was at the

university, the father of Maurice - one of my fellow students - also worked at the Ministry of Education and I heard, from Maurice, a rather amusing tale about Mr Foster. Mr Foster had to visit a convent school regularly in the course of his duties and the nuns always treated him to a very nice lunch, accompanied by a good bottle of wine - which he very much appreciated. On one occasion, however, he was going to be accompanied by his boss, a strict member of the Plymouth Brethren. He rang the convent and suggested to the Mother Superior that the bottle of wine might, advantageously, be absent on this occasion. During the course of the meal, one of the nuns appeared bearing a glass and said,
"Your usual glass of buttermilk, Mr Foster".
He had to drink it with every appearance of enjoyment although he hated the stuff - a fact of which the nuns were very well aware.

During the war, the flow of summer visitors from England and Scotland had dried up but, nevertheless, we moved from the house that my grandfather had built to a rather larger house in the centre of the town. The boarding establishment became a year-round business and we had quite a selection of interesting boarders. At one period, they included several members of the "Dublin Repertory Theatre", which had been a travelling company but settled in Bangor for a fairly long spell. They were an amusing bunch and had some great tales to tell.
One of the actors was a rather delicately bred city slicker from Dublin. On one of the company

tours, this chap had digs in a little whitewashed cottage somewhere in the West of Ireland. Feeling the calls of nature, he politely enquired the whereabouts of the toilet from the landlady.

"Just round the pile of turf, Sir."

Negotiating the pile of neatly stacked peat, he found it - a dry toilet without any door. However, needs must... He had just done the necessary when a pig ran round the pile of turf, followed by a young woman, followed in turn by an older woman. Jumping to his feet, he grabbed for his trousers but the old lady shouted,

"Don't disturb yourself, Sir, we'll catch it ourselves."

Accommodating this crowd also introduced an element of Catholicism into our Protestant, but I like to think, completely unbigoted household (largely, I suspect, due to our English mother). One Sunday, Johnny - a particularly amusing character from Dundalk - returned from mass and informed us that he had just found out the correct way to bless himself with holy water. There had been rather a lot of people trying to get to the holy water font at the church door and one old dear, much annoyed at the unceremonious jostling, modified the usual wording by saying, as she crossed herself,

"Such a bit of pushing and shoving I've never seen in all my life."

My mother capped that story with one about the Salvation Army. A friend of hers was at a meeting at which "open confession" was being practised. One lady, with great support and encouragement from the captain, was telling her story.

"Before I saw the light, I was a very bad woman and indulged far too much in drink. Isn't that right, Captain?"

"I'm afraid it was, Mary."

"And the children were not cared for properly. Isn't that right, Captain?"

"Yes, Mary."

"They weren't even fed properly and their clothes were a disgrace with holes in their socks and horrible patches. Isn't that right, Captain?"

"Indeed so."

"And as for myself, I took no pride in myself and looked a disgrace. Isn't that right, Captain?"

"I'm afraid so, Mary."

"But now I've seen the light, all that has changed. Isn't that right Captain?"

"Glory be, so it has."

"I have given up the drink and the children are properly fed and clothed. Isn't that right, Captain?"

"Yes indeed, Hallelujah!"

"Not only that but I can now take some pride in myself. I look decent for the first time in many a year and I even have lace on my knickers. Isn't that right, Captain?"

Poor man.

Stories about religious bigotry in Northern Ireland were, of course, very common. One concerned two women in a railway compartment travelling from Londonderry (Derry to the locals) to Belfast. One of them said, "What did you think of Derry?" and the other replied, "I didn't like it, it's cold and wet and full of papists". The first retorted, "Would you go to Hell where it's hot and dry and full of protestants".

Another story circulating at that time was about a Catholic priest with a small flock in a very largely Protestant area. The church badly needed redecorating but the only painter locally was a rather fanatical Orangeman. However, the alternative to using him was to employ someone from a town some distance away and, what with travelling costs, travelling time and so on, it was going to be a rather expensive job. As the Orangeman's hourly rate seemed very reasonable, the priest decided to risk asking him to do the job. On the first morning, he went into the chapel to see how things were going. Mac was painting away furiously and singing at the top of his voice "The Protestant boys are loyal and true ...", a lively and very Orange ditty to the tune of "Lillibullero".

The priest stood this for a little while and then went over to the singer, tapped him on the shoulder and said,

"Look, I don't want to start an argument or anything but I don't really think your song is very suitable for a place of worship."

Mac was immediately contrite.

"I'm sorry, Father, I didn't really think what I was singing. I always sing while I'm painting. It helps to pass the day away. But I'll stop."

"Heaven forbid," said the priest, "that I should spoil anybody's pleasure in their work but perhaps you could sing something a wee bit more suitable."

The priest went and busied himself on the other side of the church. There was a considerable delay while Mac tried to think of some alternative that would be appropriate in a Catholic church. Eventually, he thought of something which the Protestants and Catholics had in common and the meditative strains of Cardinal Newman's "Lead kindly light" filled the air. The priest looked over and observed that the paintbrush was moving in time with the very slow and mournful music. After some rapid calculations, he went over to the painter and said,

"All right, Mac, get back to the Protestant Boys."

The repertory company were approached by a local actor who said that he was desperate and if they did not accept him in the company, he would throw himself into the Lagan from Queen's Bridge in Belfast and drown himself. Not wishing to accept responsibility for any such tragedy, they agreed. After he had been with them for a short time, however, we gathered that - had he repeated the threat - they would have cheerfully given him the fare to Belfast and made sure he knew the way to the river.

During the war, the regular boarders became very much members of the family and when a reasonable number of us got together, the "craic" (to use the local expression for lively conversation) was very good. "Doing the dishes" was enlivened with "craic". My mother would invariably wash at great speed and it took three or four of us to keep up with her when drying and putting away. The dirty dishes were stacked to the right of the sink and the freshly washed ones appeared to the left. By far the largest item to be washed was a very large baking bowl and the dryers would pass this round behind my mother and back to the stack of dirty dishes. The game was to see how many times she would wash it before she caught on.

In 1942, I sat the "Senior Certificate" examination and was awarded a rather modest scholarship for the university. However, I was still only sixteen and the minimum age for entrance to the university was seventeen so on the recommendation of the school, my parents decided - and this represented a considerable sacrifice on their part - that I should stay on for a further year and try for a better grade of scholarship which would pay a bit more towards the cost of sending me to the university.

However, it was at this time that the State Bursary scheme was introduced. The basis of this scheme was that you signed an undertaking to join one of the armed services or take a particular position in industry, as required by the government, after completing a short course in engineering leading to a "wartime degree". In return, all fees were paid plus a living allowance. This was not exactly a princely sum but it represented a much better deal than any scholarship that I was likely to get. I signed up. The state bursary course was to start in September 1943 but in the meantime, another scheme appeared - called the Engineering Cadetship scheme - which I could start in January 1943 and it was suggested that I join this scheme and then transfer to the State Bursary scheme. As this paid all fees plus a weekly allowance of twenty-seven shillings and sixpence, I joined and said farewell to my schooldays.

During the period up to September, the educational side was mainly a revision of what I had already learnt at school but I did learn to do drill and fire a rifle.

Queen's University, Belfast

The state bursary course at the university did not provide a vast amount of amusement. Trying to cram most of a three-year course into two years, plus military training every Saturday morning meant a fairly strenuous week. In addition, we never had any holidays longer than two weeks and some of those had to be spent at army camps doing military training. Adding to this the fact that I was living at home and spent about two hours every day travelling to and from the university - and that I was a member of a first-aid party in the air raid precaution system - makes it obvious that student life then bore very little resemblance to the usual conception.

By the time I became a member of the first-aid party, the two air raids on Belfast (with the odd bomb falling in Bangor) were a thing of the past so that the activities of the ARP organisation in Bangor were all limited to practice events. The party I belonged to consisted of four first-aiders and a car driver. The car driver provided his own car, which he was very glad to do because it was one of the few ways that a private car owner could get petrol during the war. Our driver had quite incredibly poor sight so that in the blackout, with the restricted headlights permissible during the war, he could scarcely see a thing. The principal duty of the leader of the first-aid party was to sit beside him and give him precise instructions.

"Turn right, Mr Smith."
"Turn left, Mr Smith."
Occasionally, it was

"Stop!!!!"

That was the only excitement we had. At the end of the war, we were offered – in common with about four million other people - a medal of some sort but I did not collect it. I did not think that I had earned a medal.

Professor Warnock held the chair of mechanical engineering at the university. He was a humorous and kind-hearted man - as were, indeed, most of the engineering staff at Belfast. One morning, he had been lecturing on the relative merits of solid-injection and blast-injection diesel engines (the latter were still in use at that time) and mentioned that the solid-injection engines needed a very high-pressure fuel pump, which involved certain engineering difficulties. That afternoon, I was in the mechanical engineering laboratory and examined the old blast-injection engine there. Spotting the fuel pump on the old engine, I wondered what sort of pressure that had to generate. At that moment, Prof. Warnock walked through the laboratory. I made my apologies for intruding and raised the question of the pressure generated by the pump. Professor Warnock drew himself up to his full height.

"Young man," he said in his best professorial accent, "In any other faculty in this university, the lecturer comes into the lecture room, spreads out his notes, reads his lecture and departs. For any other information, the student must consult books and technical journals. An essential feature of university education is that the student learns how to find information for himself. Upstairs, we have an excellent technical library so my advice to you is to partake yourself thither and study the technical journals and text books until you have found the required information."

He then reverted to a strong Belfast accent and said,

"And when you've found out, you can come and tell me."

At that time, the faculty of mechanical engineering was housed in the Belfast College of Technology. Near at hand, there was a statue known universally throughout the city as "The Black Man". It may have been originally carved in black stone or perhaps decades of industrial pollution had produced the colour but at any rate it was a favourite meeting place. "I'll meet you at the Black Man" was a common arrangement. It had the additional advantage that the nearby College of Technology has a capacious porch and entrance hall which provided shelter if it rained.

There was a nice story about a Belfast man who was being visited by a cousin from Texas. He showed the cousin round the city but no matter what he saw – Parliament Building, City Hall, etc - the Texan always said that there was something much bigger and better in Texas. Eventually they came to the Black Man.

"What's that?" asked the Texan.

"That's the Black Man", said his cousin.

"You mean he was coloured?".

"Oh, no", said the Belfast man, reading the name on the statue for the first time in his life, "That's Henry Cooke".

"Why did they put up a statue to him?"

The words on the statue did not provide any very exciting information so the Belfast man answered,

"He had gallstones".

"Gee, you mean you put up a statue to him because he had gallstones?"

"Well, it was a rather special case. They took three gallstones out of that fella' and you see that plinth he's standing on? Well, they cut that out of the smallest of them."

I had a girlfriend who was studying midwifery at the maternity hospital – where else? A dance was organised there and we went along. It was a reasonably formal affair with reserved seating. On arrival, we were met by one of the nurses who were showing people to their places. She asked me my name and I said,

"Stewart".

"Doctor Stewart?" she enquired.

Being completely unaware of the subtleties of medical titles, I demurred politely,

"Mr Stewart".

She looked very impressed - and the girlfriend giggled gently.

For the benefit of those not familiar with the subtleties – some might say madness – of medical forms of address in Britain, the lower ranks of the physicians in hospitals are addressed as "Dr", even if they only have a bachelor's degree. The higher ranks, who are called *consultants*, are addressed as "Mr".

It was during my time at Queen's (as the university is always known in Belfast) that the American troops arrived in Northern Ireland. They soon appeared in the local dance hall and introduced, among other things, a rather different style of dancing. The social graces of the first wave left something to be desired and this must have filtered back to the American military authorities so that when the second wave arrived, they had been thoroughly indoctrinated as to what was considered by their officialdom to be correct social manners in Europe. In consequence, they approached the girls in the dance hall very discreetly, bowed from the waist and respectfully asked,

"May I have the honour of this dance, Miss?"

The Irish girls, not to be outdone, curtsied and replied,

"But I should be delighted, Sir" or something similar.

It did not last for long!

Not that the American soldiers were always on the receiving end of the humour. On one famous occasion, widely reported in the press to the delight of the city, a G.I. arrived at the Great Northern station in Belfast and, because it was slinging it down with rain, called a taxi and asked to be taken to the Ulster Hall (which is about two hundred yards away). The taxi driver took him for a ride all round the city and eventually deposited him at the

Ulster Hall and demanded a fare of five pounds. The G.I. took a coin from his pocket and, handing it to the taxi driver, said,

"There's half-a-crown. Now hop it before I call a cop. I was born and bred in Belfast."

A minor bit of bad luck, associated with earlier family history, also occurred at this time. My father had an uncle who had emigrated to the States a long time earlier - probably about the turn of the century. The last the family heard from him was a letter in which he described himself as a horse dealer. During the war, however, a couple of American soldiers in transit came to Bangor and enquired if anyone knew anything about a Tommy Stewart who had lived in Church Street. Unfortunately, the person they asked was not an old Bangorian and simply replied in the negative. My father did not hear about it until much later and by then it was too late to do anything about it.

Returning to educational matters and skipping on to the late spring of 1945, the state bursars were coming near the end of their course and a military officer from the Ministry of Scientific and Technical Manpower come to interview us. He lined us up and addressed us somewhat as follows in a delightfully clipped military accent: -

"Now then, you chaps, you have a form to fill in and you will see that you can state a preference for the Army, the Navy, the Air Force or industry. I should not, of course, wish to influence your free choice in this matter but I should mention that should you state a preference for industry, I can guarantee you all jolly good jobs within a few weeks. Should you opt for one of the services, however, you may find yourselves hanging around for a year or so." Faced with this free choice, we all opted for industry - with the exception of two blokes from old naval families. They put themselves down for the Navy and were kept hanging around for over a year.

(During the course of my subsequent career as an engineer, I met erstwhile state bursars from other universities. In 1945, Edinburgh received exactly the same treatment as Belfast whereas the Glasgow contingent were all "volunteered" for the army. A useful object lesson on the ways of officialdom but I suppose it saved someone a lot of work.)

After this preliminary canter, Major Walters invited us into an office for individual interviews. After some waffle about whether I played rugger and had been a house captain, he eventually cleared his throat and said,

"Now, engineering, what sort of engineering interests you?"

Having done my training at Belfast with its shipbuilding tradition, I had absolutely no doubt.

"Steam turbines and heavy diesels, Sir."

"Jolly good," he said, "Rolls-Royce."

I suppose the various jobs available had been allocated by using a pin to pick us out from a list. Still, young people fresh from university do not usually have much of a clue about what really suits them so I suppose the pin method is as good as any.

So Major Walters terminated any possible military career - with one small exception. During the first year of the State Bursary course, we had done infantry training in the university Senior Training Corps (I rose to the dizzy rank of lance-corporal) and in the second year we had signals training (I was demoted to private, it was my friend Billy's turn to be lance-corporal). The signals CO was a very decent old bloke. (Well, he seemed old to us then!) For that final year he had made all the arrangements for us to have a fortnight's session at the College of Signals at Catterick and now he had no soldiers. Would we back him up and go anyway? So we did. It was wonderful. The weather was terrific for the whole fortnight. We were accommodated in some of the army's most modern barracks. The war was over in Europe and the instructors who had put our predecessors through a fair approximation to hell to get them ready for the rigours of war thought they ought to make up for it by being nice to us. We even fed in the sergeants' mess! The exercises seemed to involve a series of stunts that took us in Jeeps through beautiful Yorkshire countryside. We did lay a few field telephone lines, but nothing too strenuous.

Why should we have had such a good time? Life is really most unfair. We were on the winning side but so many of our predecessors were on the losing side. Omar Khayyam again:

"The Ball no Question makes of Ayes and Noes,
But Right or Left as strikes the Player goes;
And He that toss'd Thee down into the Field.
He knows about it all – He knows – HE knows!"

Before leaving matters military altogether, I should mention another example of the lack of justice in life. There was one real soldier in the family - my elder sister, who was in the ATS. She was stationed at an anti-aircraft battery (where she met her husband but that is irrelevant to the present story). One of the soldiers in the unit was from the Irish Republic and, going home on leave on one occasion, he did not bother to come back. There was nothing the CO could do about it. Eire was neutral. After about a year, however, the CO got a letter from this character saying that he was unemployed and a bit fed up. What would happen to him if he came back? King's Regulations provided no guidance as to how to write back to a deserter in a neutral country! He consulted higher authority. The query went quite a long way up the line. Eventually, a question came back by telephone. What sort of a soldier was the man? Answer: Absolutely first class until he disappeared. The CO was then told to write to the chap in his own fair hand, without keeping any copies, telling him to spin a good story and it would be all right. (There was a war on and good soldiers were not to be sneezed at!) Patrick obliged with a very convincing letter about a mother at death's door and various other complications, was granted retrospective compassionate leave and returned to the unit.

    Everyone was quite happy until the Army Pay Corps caught up with events and coughed up a year's back pay! The battery almost mutinied.

Rolls-Royce - first stint

So my request for a Rolls-Royce (at a cost of half-a-crown) from the butcher in Bangor was not quite granted. I did not get a Rolls-Royce but at least I went to work for the company - on aircraft engines, as it happened, not motor cars.

I had been invited over to Derby to be interviewed by the company early in the summer. In the waiting room, while waiting to be "done", I found myself in the company of a chap from Bangor in Wales. As neither of us had ever met anyone from the "other" Bangor, we had quite an interesting chat before I was called in for my interview with the personnel officer, Arthur Livesey. During the preliminaries, I mentioned the odd coincidence of two people from the different Bangors meeting in Derby. "Oh dear," he said, "I thought you were from the same place and could travel together."

My employment at Rolls-Royce actually started in August, just after VJ day.

I travelled over by the Larne-Stranraer ferry and had been informed that I would probably have to make at least two changes of train before I got to Derby. However, when I got to Stranraer, I enquired of a railway porter which train I should catch to get to Derby. He pointed to a train and said,

"Get onto that one and keep your mouth shut."

When I got on the train, I rapidly discovered that I was on a troop train but one which, in fact, was travelling direct to Derby. Once the train was going, I confessed to the service chaps in the compartment that I was really there under false pretences. They were fairly amused. It being a troop train, there were no ticket inspectors or anything like that to cause trouble. At one stop, they all got out to go to the NAAFI which, of course, I could not do and there was no other possibility of getting refreshments. However, the chaps brought me back a mug of tea and some food from the NAAFI. So I got to Derby comparatively painlessly and staggered off the train at about 7 o'clock in the morning.

My stint at Rolls-Royce commenced with a course of practical training, which lasted about six months. The first two months were spent in the training workshop at the Derby Technical College, where I became very friendly with one of the instructors - Charley Moughton. Charley had started life in one of the skilled trades but had then become an instructor in the college and had, if that is the right expression, worked his way up in the world. At any rate, he was one of the best-read people I have ever met. At his home one evening, he told me that neither he nor his wife smoked or drank and his sons were working so really they were reasonably well off. Their only extravagance was a really good holiday every year. He would like to go abroad and feel that, for a fortnight, he was something of the "English Milord". Going abroad was a bit difficult just after the war but there was no difficulty about going to Ireland, so could I recommend a good hotel in Dublin. I obliged and, of course, saw him after his holiday to ask how it

had gone. He had had a wonderful time and, so he said, been quite overwhelmed by the charm of the locals. In particular, he described his visit to the "Sixteen Room", which commemorates the 1916 rebellion, in the National Museum. After examining the various exhibits - old uniforms with bullet holes, last letters by leaders of the rebellion who were about to be executed, etc, he approached the caretaker and told him in his best cultured English that "as an Englishman, he had been greatly affected by the exhibition. It had given him great food for thought." The caretaker took him by the lapel of his jacket and replied, in a strong Dublin accent,
"Ah well now, all these things are all very fine and large but what's the good of them to working men like you and me!"
On another occasion, he asked directions from a man who was working up a ladder. The workman obliged with very clear directions in a rather loud voice and then came down the ladder.
"Come on and I'll show you."
"If you were going to show me, why did you give me all those instructions?"
"Sure, that was for the foreman's benefit. He's inside."

The next manufacturing area included in my practical training was the pattern shop, where the shop steward was also secretary of the Rolls-Royce Rugby Club. He came to find out if I played rugby and I told him that I did although, in fact, I was no great shakes at the game. However, I rapidly discovered that in the Derby area - at that time - the knowledge that rugby was played with a ball that was not round was sufficient to rank me as an experienced player. So I found myself turning out for the Rolls-Royce Second XV. Rugby at that level was more amusing than skilful and of the type so wonderfully described by Michael Green in "The Art of Coarse Rugby".

After a peregrination through various manufacturing centres, my last port of call (before starting to work in earnest) was the compressor test bed. One of the compressors being tested was the AJ 65. It was the first axial compressor to be built by Rolls-Royce and – as might be expected with something revolutionary – it initially gave its fair share of troubles (although, of course, the company got it right and it was very successful in due course as the compressor in the Avon engine). Unlike the centrifugal compressors that preceded it, the axial compressor was very intolerant of the "surge" phenomenon (which occurs when the compressor is subjected to excessive back pressure) and frequently chucked all or most of its blades out through the air inlet (not the outlet, which you might expect). It was the job of the test bed labourer to sweep them up. On one occasion a very senior member of the Rolls-Royce management was showing an Air Vice-Marshall round the test beds and they came to the bed where the AJ 65 was located. The senior executive explained that this new variety of compressor had a multitude of small blades instead of the large single impeller in a centrifugal compressor.

"Very interesting," said the visitor, "Tell me, how many of these blades would there be in such a machine."

The senior executive had no precise idea, looked around and asked the only person in sight - the test bed labourer.

"How many blades are there in that compressor?"

The test bed labourer, an old soldier, came smartly to attention (complete with sweeping brush).

"Three buckets full, Sir."

In the spring of 1946, I became a technical assistant in the compressor office. There were about sixteen of us working there but two in particular come to mind - Frank and Taffy. Frank was an incredibly orderly man in everything he did whereas Taffy - a Welshman - tended rather to the "lightning genius of the Celt". Frank bought the Manchester Guardian every day and read it at lunchtime while eating his sandwiches. When he was finished with it, he passed it on to Taffy. The lunch break was not quite long enough to finish the paper completely - Frank read it from the first word to the last. So, in his orderly fashion, he would finish the preceding day's paper before he started the current one. Needless to say, he got further and further behind until he was eventually one day out of phase. At that point, Frank was still buying the paper every day but Taffy was reading today's paper while Frank was reading yesterday's!

It was Frank who sold me my first motorbike - a 500 cc TT replica Triumph (1929 vintage – what would it be worth now if I had held on to it?). Intrinsically of historical interest, it had been made more interesting by various additions. During the war, an undamaged engine from a Messerschmitt had fallen into the hands of the British and this was tested by Rolls-Royce to see what they were up against. Associated with it was a lot of redundant wiring and Frank had used this to rewire the old Triumph. It had other interesting modifications - which I only appreciated later.

In the spring of 1946, before I acquired the motorbike, I mounted my pushbike and went youth hostelling in Derbyshire. Despite the very intensive State Bursary programme at Queen's, we had managed to do some hostelling, cycling down to the Mourne Mountains on Saturday afternoon - after the morning's military training - hiking through the hills on Sunday morning and afternoon and cycling back to Bangor on Sunday evening. Such energy we had - sic transit gloria... On longer breaks, we cycled across the border into the Republic where we could enjoy the luxury of street lights that were lit and - luxury of luxury – ice cream, which was unobtainable in the United Kingdom during the war.

    On my first expedition into Derbyshire, I was heading for Ilam youth hostel and was not sure if I was on the right road. I caught up with some hikers and asked the way. My Ulster accent must still have been very strong because I was told the time. However, after that small difficulty, conversation developed and it appeared that we were all going to the same place. One of the hikers was a girl called Betty who became my regular girlfriend and later my wife. Also in the party was Betty's cousin Mick who, when he heard about our forthcoming marriage a couple of years later, said,

    "You're not going to marry that Irishman, are you? "

    (Mind you, that was comparatively mild. One of Betty's friends said, "You're not going to marry that Irish peasant, are you?") I can hardly have made a very favourable impression! During the same weekend, I met other "hostellers" who remained friends over the years and, a final bonus, a fellow state bursar who had been "volunteered" for a job in Stafford also turned up at the youth hostel. It was a good weekend.

Thus provided with an interesting job, friends and a girlfriend, life at Derby became really very pleasant, despite rationing and the other shortcomings of the immediate post-war period. With the return of the rugby season in the autumn, hiking and cycling were abandoned because every Saturday afternoon was booked. This caused some comment from the girlfriend to the effect that I was now invariably missing on Saturdays.

"Why not come and support the matches?" I said. "Even if it is an away match, there is always plenty of room at the back of the bus."

"Oh no," she said, "Rugby is a rough game."

"Nonsense," I said, "In fact, the statistics show that there are more serious injuries at soccer than there are at rugby."

Eventually, I persuaded her to come to the club's seven-a-side competition at the end of the 1946/47 season. I must have been on the field for almost a minute before I broke my collarbone. Somehow, my statistical arguments did not carry the same weight after that!

The summer came and went and, in the autumn, I had completed two years at Rolls-Royce and this, on the basis of some abstruse calculations, meant that I had done my stint in return for the state bursary course and could now return to Belfast University to finish off my degree course. So it was farewell to Rolls-Royce and Derby.

Back to Queen's

Finishing off the degree course involved another pretty intensive year because, in addition to repeating all the final year subjects, we had to make good all the earlier years' subjects that had been skipped during the state bursary course.

Nevertheless, life at the university was beginning to take on a more normal aspect, with numerous balls and other events. In particular, there were some functions at which evening dress was de rigueur. That was rather a problem; particularly as clothes rationing was still in existence. However, an expenditure of £5 (quite a lot of money in those days) at a second-hand clothes shop solved the problem.

After one ball, the last formal university function I attended, I had been offered a bed for the night in the ground floor flat of some friends - John and Jean were the couple's names. Arriving at the flat, all was dark but there was a note under the knocker.

"The window beside you is open, climb through it and the bed is immediately in front of you. Jean."

I stuffed the note in the pocket of my dinner jacket, climbed through the window, found the bed and had a good night's sleep. As it happened, I never had occasion to wear my dinner jacket again for many years. By then, I had been married for several years to Betty, the English girlfriend mentioned above. As a good wife, she nobly decided to press my dinner jacket for the important occasion that had arisen. She found the note.

"Who," she asked, "is Jean?"

As some years had elapsed, I was initially somewhat nonplussed but then I remembered and told her all about my friends John and Jean. She appeared quite happy with the explanation but less trusting friends thought it was quite a good story to have thought up on the spur of the moment!

Returning to the university for the moment, I was recruited in that final year by British Thompson-Houston representatives who were on the "milk round". (For the benefit of those not "in the know", the milk round is an operation carried out by large firms at times when particular graduate skills are in short supply. Representatives of the firms visit the universities and interview students, offering jobs to likely characters before they actually do their finals. At a later period in my career, I myself participated frequently in the "milk round" and after the "Troubles" had started in Northern Ireland, I had a monopoly of the Rolls-Royce "milk round" to the Irish universities for a couple of years.) Reverting to BTH, the arrangement was that I would do another year of practical training as a graduate apprentice with a view to making a career in the company. The final examinations were in June and as I was not due to start with BTH until September, I had two whole months of freedom in "Bangor by the sea".

While still at Queen's, it became obvious to me that I could have stayed on and added an electrical engineering qualification to my mechanical engineering qualification - all at the government's expense and without any difficulty.

The contrast with the effort necessary to obtain my first scholarship to the grammar school (worth all of fourteen pounds per annum), which had involved frequently working till midnight at the age of ten, was a bit striking. The difference was that now I was "on the books" at the Ministry. I was beginning to learn something about the way of the world. Fortunately, I was dissuaded by the Advisor of Studies who told me to get out into the real world where there was much more to be learnt.

Almost immediately after the last examination, I took the night steamer to Glasgow, where I met Betty, who had come up on the night train from Derby. From there, we set off for a fortnight's hiking (and some hitch-hiking) in the Scottish Highlands, staying at youth hostels with wonderfully romantic names like Loch Lomond, Crianlarich, Glencoe, Ratagan and Kyle of Lochalsh. Time ran out so we did not manage to go "over the sea to Skye." One of our stops was at Ratagan hostel where we made the acquaintance of the warden - Dominic Capaldi, a man remembered with affection by a generation of youth hostellers. Despite his name, he was very much a Scot and always wore the kilt (although I do not suppose it can have been the Capaldi tartan!). The hostel was small and his own accommodation was very limited so he used to eat his meals in the same dining area as the hostellers. This gave rise to an odd little quirk of his. He would be sitting eating a meal in the late afternoon when hikers would arrive. Seeing him sitting there, the new arrivals would go to him and enquire,

"Where can we find the warden?"

He would reply very solemnly, "the warden is having his tea", so they would find a seat and wait patiently for the warden to appear.

Those of us who stayed more than one night in the hostel would eagerly look forward to seeing other people getting caught.

Also staying at Ratagan were an English clergyman and his wife. We got on very well with them and he was all for taking advantage of Scottish law (as he understood it) and marrying Betty and me on the spot. We thanked him but declined. Apart from the fact that he seemed to be rushing things a bit, I was dubious of his interpretation of Scottish law and Betty thought that she should perhaps mention the matter to her parents.

At the end of a fortnight, Betty's holiday was over so she returned to Derby and I returned to Bangor where, about that time, a rather nice example of my father's rather wicked sense of humour occurred. The town council had decided that we ought to have a planning officer and had employed a suitable candidate. This gentleman gave a lecture on the subject of town planning and my parents and I went along. The lecture was mainly devoted to the evils of ribbon development and there was a period for questions and discussion after the lecture. My father got up.

"You have been talking at some length about the evils of ribbon development. Now recently, I was driving my wee car from Belfast to Ballymena and, as you probably know, there aren't many houses at all along that road and, as I was driving along, I was asking myself what would happen if the car should break down. I would be absolutely stuck. Now, if there had been a nice bit of ribbon development along that road, I wouldn't have been worried at all."

The lecturer looked rather nonplussed.

Memory is a strange thing. No doubt if I consulted the records, I would find that there was the usual amount of rain during that long two-month's holiday in Bangor but it seems as if the sun shone every day and that life was one long succession of swimming, walking round the coast to Groomsport or Helens Bay and enjoying the company of my friends in the evenings. Omar Khayyam yet again:

> "Alas that spring should vanish with the rose
> And youth's sweet-scented manuscript should close.
> The nightingale that in the branches sang,
> Ah whence, and whither flown again, who knows."

However, all good things come to an end and the time came to say good-bye to family and friends and head for Rugby.

British Thompson-Houston

The graduate apprenticeship only lasted a few weeks because by the time I arrived at BTH, two of the chaps I had worked with at Rolls-Royce were working there in the gas turbine department. They wanted someone with some gas turbine experience to work on the testing of the components of BTH's new marine gas turbine.

One character I came across at BTH had the interesting habit of always answering the phone with "Mr Jones" - not simply with "Jones" or his telephone extension number or something else appropriate. This seemed odd to me and I asked colleagues about it. The resulting story is an interesting reflection on English social and industrial conventions. Before the war, this chap had been a fitter at BTH. He was called up in due course and served in Burma (I think) under conditions where the causality rate was very high and promotion correspondingly rapid. When he was demobilised at the end of the war, he was a full colonel. In due course, he turned up at BTH and - as was his right - demanded his job back. Needless to say there was a form to be completed, which included a request for information on his army record. When this form came to the notice of the BTH top brass, they were horrified to find that a full colonel was asking for his job back as a fitter! His attitude was very simple; he had done what he had to do during the war, now it was over and he wanted to get on with his life! They wanted to provide him with some sinecure of a job appropriate to a retired colonel but he would have none of it. Eventually a sensible compromise was reached and he became a supervisor in an appropriate department in which both his practical skills and his obvious leadership gifts could be utilised. So far so good, but people insisted on addressing him as "Colonel Jones". Nothing doing - the war was over and he objected strongly to being reminded of it - which was why he always answered the phone with "Mr Jones."

Being back in digs after a year spent at home did not prove very attractive. Apart from that, I was travelling up to Derby every weekend to see Betty, who had been my regular girlfriend for over two years. So I popped the question. Betty thought she could put up with me and her mother seemed keen on the idea but, of course, I should have to ask her father. Betty was only twenty. Everyone else was tactfully removed from the dining kitchen where Mr Anthony was reading one of his favourite cowboy stories.

Nervous cough.

"Mr Anthony."

Mr Anthony carefully marked his place on the page with his finger and looked up.

"Yes?"

"Well, er, Betty and I would like to get married."

"If that is what you want, I suppose it will be all right," said he, returning quickly to his book before he lost his place.

Any illusions I may have had about a quiet wedding were rapidly shattered. Mrs Anthony and Betty's five sisters saw to that. I only made one modest suggestion about the arrangements. As we were going to be crossing the sea to Ireland on our wedding night I suggested that one of the hymns might be "Eternal father, strong to save" which includes the chorus line "for those in peril on the sea." I was overruled. As Betty was the first of six daughters to be successfully married off, the Anthony family insisted on "Now thank we all our God."

We were lucky enough to start our married life in a very nice flat, which was part of a farmhouse about nine miles from Rugby. It was an ideal place to start our married life and the old lady who owned the farm could not have been nicer.

We started to attend the village church and, in the process, caused the attendance to rise by a very substantial percentage. When preaching his sermon, the vicar would keep his eyes fixed firmly on the ceiling of the church, perhaps to keep his thoughts fixed on heavenly things or, more probably, to avoid facing the reality that he was very often preaching to Betty and myself plus his wife and the verger. When prayers were being read, the verger would try to help matters by responding with an incredibly loud AMEN.

It all reminded me of the story about Dean Swift (the Gulliver's Travels bloke) who, at one point in his career, had a parish in the West of Ireland. At that time, the Church of Ireland was the Established Church, even though only a fairly small proportion of the population were Anglicans. In consequence, churches were maintained in parishes where there were no Anglicans at all. Swift had one of these parishes and the only people at the services would be himself and the verger – both of whom were paid to be there. In consequence, he would amend the services appropriately.

"Dearly beloved brethren, the scripture moveth us in sundry places to acknowledge and confess our manifold sins and wickedness ..." would be replaced by

"Dearly beloved George, the scripture moveth thou and I in sundry places to acknowledge and confess our manifold sins and wickedness ...".

The stay on the farm gave me a new insight into country people's views on foxhunting. Whenever there was due to be a hunt in the vicinity, any able-bodied people available were mobilised to anticipate the arrival of the fox and deflect it onto somebody else's land so as to minimise the damage caused by the hounds and the horses.

Our landlady's brother had his own way of dealing with foxes. He would go out late at night with his gun and utilise his detailed knowledge of some woods in the vicinity to take up a position where a fox was likely to appear. I went out with him a couple of times and was amazed at the way he could remain absolutely motionless and without making the slightest sound for what seemed to be incredibly long periods of time. If the fox did appear, he would despatch it quickly and accurately.

This was not a question of sport. They were raising sheep and foxes were a genuine menace.

My means of transport between the farm and work was the old Triumph but it became obvious that the main bearing was on its last legs (if you will excuse the rather odd metaphor). I stripped the engine down, removed the bearing and gave it to a colleague at work who had reason to visit the bearing agency at the weekend and undertook to get me a replacement. He handed over the bearing at the counter and the man he had given it to returned after a few minutes looking rather puzzled.

"May I ask," he said, "where you got this bearing?"

"It is from my friend's motor bike," said my colleague.

"Your friend," said the other, "must have a very powerful motor bike. This bearing came out of a Rolls-Royce Merlin aircraft engine!"

Our residence in the flat on the farm did not continue very long because I rapidly came to the conclusion that my prospects at BTH were not very bright and when an opportunity arrived to take a more responsible job, we moved to Lincoln.

Ruston and Hornsby

In the Gas Turbine Department of Ruston and Hornsby at Lincoln, I was the Project Engineer for a scheme which - in retrospect - seems quite incredible although it must have seemed sufficiently sensible at the time to justify money and effort being spent on it. It was one of a group of projects sponsored by the Ministry of Fuel and Power and these projects involved several companies who were supposed to interchange information arising from the research and development work. It all provided excellent experience for a young engineer - starting a project from scratch, negotiating and exchanging information with the ministry and other firms - and leading a small staff, about whom more anon. My particular job concerned the combustion of peat on an industrial scale.

Finding suitable living accommodation proved difficult and what we found was rather a comedown after our excellent flat on the farm. We started off in rooms, not what we wanted but accommodation was difficult just after the war. The man of the house was a driving instructor but had previously been a bus driver. On one occasion, he took an excursion to London and, not knowing the city very well, he stopped and asked directions from a policeman who pointed to a London Transport bus and said:

"Just follow that bus and you will get there."

The result was that the Lincoln company's bus ended up in a London Transport bus depot - to the loud applause of the London Transport personnel present.

After the rooms with the driving instructor, we moved to a caravan owned by a vicar with a parish on the outskirts of Lincoln. He was an interesting character who had once been a missionary in Canada, about which he told some odd tales.

Apparently there is a variant of the Pacific salmon that looks very similar on the outside but has white flesh instead of pink flesh. The first operation at the canneries was to nick the skin of the salmon and any with white flesh were simply rejected. Two chaps offered to buy these reject salmon from the big canneries (at a very low price, of course) and they then set up their own cannery to deal with them. They stuck a label on the cans which said:

"PURE SALMON. GUARANTEED NOT TO TURN PINK IN THE TIN."

They were sued by the other canning companies. The case eventually ended in the Canadian High Court, where they lost and had to remove the offending label.

He also claimed that there was once a law in some part of Canada whereby a couple found sleeping together were deemed to be married. Not only that but if one of them was already married, the offender could be tried and jailed for bigamy![*]

---

[*] (My youngest son now lives in Canada and has tried without success to confirm these stories but, since they came from a clergyman, surely they must be true!)

Describing religious life in the Far North, he said that most denominations were to be found but Baptists were rather rare, total immersion having a distinctly limited appeal North of the Arctic Circle.

Most of the staff of the Gas Turbine Department at Lincoln were fairly young and came from other parts of the country. Recreation to our taste was somewhat lacking so a few of us formed a square dance band to brighten things up. Because I was unable to play any instrument, I became the caller. After a few practices, we decided the time had come to hold our first dance and we got permission to hold it in the canteen. We went to some trouble to decorate the room suitably and lots of the staff and their wives and friends turned up.

Everything seemed to be going well until we struck up for the first dance. We had not, however, reckoned with the age of the building and the fact that the dust of at least a century had accumulated between the floorboards. To give it its due, the old floor must have been well sprung because, after a few bars, the enthusiastic battering on the floor released the dust that had accumulated over the decades. Everyone was soon coughing and spluttering and it was almost impossible to see across the room.

However, we opened all the doors and windows and the dust gradually dispersed. In the end, it was quite a successful evening.

Eventually, we played for quite a few square dances around the area and even, if I remember right, at a Scots function - although I do not remember anyone actually paying us!

We did not live in the vicar's caravan for very long because, as a key worker, I was allocated an aluminium prefabricated bungalow by the local authority. In the meantime, however, Betty had become pregnant and returned temporarily to her mother's house until the bungalow became available. It was during this period that our first child arrived. She had the baby boy in a nursing home in Derby. Needless to say, I hastened to Derby to see her and my son and heir. I had been given instructions to go up Queen's Street and straight on at the "Five Lamps"; I would find the nursing home on the left. I never found the "Five Lamps" but arrived at the nursing home without difficulty. After spending some time with them - they were both well, thank goodness - I emerged from the nursing home determined to find the "Five Lamps" and have a drink to celebrate the important event. Coming to a rather complicated road junction, which I had passed on the way to the nursing home, I asked a passer-by where I could find the "Five Lamps".

"There they are," he said pointing to the lamps around the road junction, "One, two, three, four, five."

(The story has a sequel. Some years later, I was driving through Naas in Co. Kildare in Ireland with Betty and James, who was then a boy of about three. I suddenly stopped the car. There, at the side of the road, was a pub called "The Five Lamps".

"I am afraid you will have to wait for a minute." I said as I went in to have my long overdue beer.)

Before very long, we moved into the aluminium bungalow. It had a small garden, if that is the right word. I was convinced that the material for making the roads in the estate had been piled on the bit of ground around my bungalow - and that they had probably mixed the concrete there as well. Since most of the other men on the estate were of the same opinion, however, I can only assume that we had landed on a particularly hard bit of geological strata. Be that as it may, we struggled to dig it and most of us soon had a fair pile of stones somewhere on our "garden".

One Friday, the council workers came to deal with a small area on the estate which was intended to be a green. Despite having suitable machinery, they spent the whole day preparing the area for the planting of grass and, by the end of the afternoon; they had assembled quite a nice little pile of stones. On leaving, they mentioned to one of the residents that they would send a pick-up truck on Monday morning to remove the stones. This piece of intelligence spread rapidly round the estate and everyone spent the weekend carting barrow-loads of stones to the green.

When the council workers arrived with their pick-up truck on the Monday morning, they found that the green was covered by a miniature mountain of stones. After some discussion, the pick-up truck and the council workers disappeared and, after some time, a large lorry appeared but even this had to make several journeys before the pile of stones finally vanished and work on the green could resume.

At work, I had a small design office with a staff consisting of two Poles, one Swiss, one Scotsman and two Englishman, one of whom was an Oxbridge character. In due course, we were joined by Harry. Harry had been a turner and had had some sort of accident at work, which made it impossible for him to pursue his normal occupation, so he had to be found an office job. He was allocated to our department for general duties. On his first day with us, he opened the office door, gazed around at the assembled staff proudly operating their slide rules (we really did use them in those days!) and draughting their magnificent schemes. He paused for a moment and sighed,

"I never thought I should come down to this."

Fortunately, he soon decided that we were tolerably human.

As might be expected with such a mixture of nationalities in the office, conversation sometimes turned to questions of semantics and, in particular, dialect. On one occasion, the Oxbridge chap pontificated.

"The dialect word I really dislike is the word `mash' they use for making tea in the Midlands. I can just imagine them squeezing the drops of nectar out of the little muslin bags."

Harry did not seem to have heard of nectar but he quickly voiced his suspicions,

"Na then, lad, there's no need to be filthy."

Harry was a regular supporter of Lincoln City football club. One Monday morning he told us about an experience at the match the previous Saturday. He had blacked another supporter's eye and a policeman was called. When asked to explain his action, he described how he had only intervened to stop the bloke with the black eye from battering a little chap of very diminutive stature.

"Is that true," the policeman asked the man with the black eye.

"Well yes, I suppose it is," he said.

The policeman turned to Harry,

"I should black the other one if I were you."

Real football hooliganism was practically unknown in those days. Perhaps there is a connection!

When Christian Meyer, the Swiss designer, joined the office at Ruston and Hornsby, his English was still somewhat scanty so we made an arrangement - he taught me German and I taught him English. As practically no one then had a car, we both had to spend quite a long time on the bus getting to and from work, so we devoted that period - along with a lot of other spare time - to our linguistic studies. We both went to night school as well. The oral part of the learning curve was very much a matter between the two of us, both at work and elsewhere. This very intensive arrangement helped us both enormously and after about nine months, I was sufficiently "over the hump" to be able to read German newspapers and books with some pleasure. After that, improving my German became easy. Christian's experience must have been similar because after about a year, he came into the office one morning looking as pleased as Punch. When we enquired the reason for his pleasure, he told us that he had been at a dance the night before and the girl he was dancing with had asked him,

"What part of Ireland do you come from?"

I do not think I can have done as well as Chris because I was never accused of having a Swiss accent when I was living in Germany many years later.

Another of the designers was Bill, who also played the drums in our square-dance band. Bill had done his national service in the Navy and although he only talked in general terms about the character-building benefits of life in the senior service, we all had visions of him heroically battling with arctic gales and so on. It was only after some time that we discovered that his entire national service had been spent in a shore-based job at

Scapa Flow and that the only time he had been to sea was on the ferry crossing to and from Thurso.

Apart from the Project Office, which I ran, there was a main gas turbine design office led by Bill Sharpe. One of the designers in that office was a gritty little Yorkshireman called Derek Green who had suffered from poliomyelitis and was still walking with the aid of sticks when he first joined the company. During the period that I knew him, he struggled bravely with his disability and - I am glad to say - overcame it to a very large extent. He had a very dry sense of humour. He lived on the same road as Bill Sharpe but further out from the centre of the city. One day, several of the more senior members of staff were in Bill Sharpe's office discussing the important matter of time-keeping. Derek came into the office for some reason or another. He listened to our discussion for some time and then chipped in,

"Yes, Mr Sharpe is very keen on time-keeping. He looks up from his breakfast every morning to make sure I am on the bus."

The job involved some travelling around the country and, indeed, one trip to Holland.

On an expedition to Scotland, I met a fairly senior engineer of the North of Scotland Hydroelectric Board. The dams were built in the scantily populated Highlands of Scotland but most of the electricity was to be sent south. The people in the Highlands were, naturally, not unduly keen on the project and, as a sweetener, it was agreed that electricity had to be supplied to every residence in the Highlands – no matter how small or remote.

The hydroelectric board engineer told me about a first quarterly inspection of the meter in one remote croft where the chap reading the meter noted, to his surprise, that the meter had hardly moved at all. So he said to the old lady who lived there, "You don't use much electricity, ma'am, do you?"

"Och, no," she replied, "I just put it on while I light the lamps."

An expedition to Thurso in the far North of Scotland provided me with a nice story, although I have since been told by Scots friends that the same story is told in many versions about practically every town in the country. The story concerns a visit to Thurso by Victor Hugo, the famous French novelist. Being an important man, he was invited to stay with the Provost. On the Sunday morning, they were going to church together and as it was a fairly brisk sort of morning, Victor Hugo was striding out and enjoying the air. The Provost spoke to him,
"Excuse me, Mr Hugo, would you mind shortening your step. I would'na like the folk to think I was enjoying a walk on the Sabbath."

An obvious technical visit was to the Republic of Ireland where peat was already being burnt on an industrial scale. I visited the Bord na Mona research establishment and a peat-fired power station as well as a site where peat was being harvested on an industrial scale. Another man I visited was Frank, who - although he had spent many years on research into the combustion of peat - was now the Chief Engineer of a distillery. He showed me round the distillery, which included an enormous steam engine. It had, I was informed, been bought second-hand during the nineteenth century. It really looked like something out of a museum and developed, apparently, all of ten horsepower. I expressed my surprise, pointing out that a little electric motor could perform the same duty, take up a lot less space and require practically no maintenance. Frank smiled. He had of course told the directors of the distillery the same thing but they were adamant that no single item in the distillery could be altered in any way because, in the first place, it might just possibly alter the flavour of the whiskey and, secondly, all the joints were sealed by the Customs and Excise to ensure that no whiskey could get out of the system without duty having been paid.

Frank did, however, tell me that local rumour suggested that one exception had occurred to this rule. When the Republic of Ireland became independent from Great Britain, there was an interim period before the Irish customs and excise authorities took over from the British authorities. According to the rumour, after the Irish authorities did take over, there was a tap giving pure whiskey in a house on the other side of the street.

The four years we spent in Lincoln were very pleasant in many respects and extremely educational from an engineering point of view. However, a change began to look desirable for various reasons. In the first place, our second child was born with a heart defect that resulted in his death after a few months. Nowadays, an operation would probably be possible but in those days, the possibility was only a vague rumour of something happening in the United States. In addition, the job was not developing in the way I had been hoping and it seemed to me, perhaps quite wrongly, that the internal politics were developing against me. Perhaps most important, however, was the fact that - although no economist - it appeared to me that the whole project was quite ill-conceived and a waste of taxpayers' money. The job had, however, provided me with contacts in the Irish Republic and as a result of these, I was offered a job with a small firm in Carlow, which is about fifty miles south-west of Dublin. As I had never worked for a small firm and the terms and conditions offered were quite generous, Betty and I decided that the change would do us both good so I accepted the job.

Thomas Thompson & Sons Limited, Carlow

1953 found us heading for the Republic of Ireland. We initially left our two-year old son James with his grandmother in England while we went across the Irish Sea to get a home together. The company in Ireland provided us with a pleasant house on the outskirts of Carlow with just one field between us and the River Barrow. Civilisation appeared. We had a fridge for the first time in our lives and there was a station wagon that went with the job. Not only that, we had left food rationing behind in England and with no rationing in Ireland, life really did look extremely rosy. It is hard to imagine nowadays but Betty found her first visit to the butcher's shop in Carlow an absolute revelation. She ordered something approximating to the then English meat ration for a week and was astonished to see the other customers carrying out large parcels of meat.

It took about a fortnight for us to get things more or less organised in Carlow and then Betty went back to Derby to collect our son. She returned with him on the night steamer and I drove up to Dublin to meet the steamer at the North Wall, where it docked at that time. To get to Carlow, I had to drive for quite a distance along the River Liffey, over which there are many bridges. Each bridge forms a quite important traffic junction and there was a Civic Guard on point duty at each one - I suppose they have long since been replaced by traffic lights. However, at one of the bridges, the

Civic Guard put up his hand to stop me and then beckoned me over beside him. Wondering what I had done wrong, I drove up to him and put down the window. He stuck his head in and said,

"Are you going up as far as the next bridge?"

"I am," said I.

"Would you mind", said the Guard, "telling the Guard up there that the Inspector is on his way up?"

At that point, Betty began to realise that life in Ireland was somewhat different from that in England.

The social life in Carlow was perhaps the best that I have ever known. Things have probably changed in the last forty years, but at that time the combination of being a Protestant and having a reasonably good job ensured a highly favoured position (much more so than in Northern Ireland, strangely enough). We were quite astounded to be invited to a hunt ball in a nearby aristocratic residence soon after our arrival. That was quite an eye-opener. The ball started round about 8 pm and was interrupted for the evening meal at about 11 pm. After that, we danced on until 4 or 5 in the morning when hot soup was dished out as a "corpse reviver" before we set off to drive home. The hunt balls were usually held on a Friday evening to avoid dancing through into Sunday. Most companies worked on Saturday morning in Southern Ireland at that time so we used to go home, change from our glad rags into working attire and doze for a couple of hours in a chair (we dare not go to bed in case we should really go to sleep). We then had some breakfast before going to work. The real hunting fraternity, on the other hand, went home and changed out of their dinner jackets into hunting attire, had breakfast and then set off for a day's hunting on horseback. Not for Alex!

In retrospect, the move from England to southern Ireland seems more like a change in time rather than a change in place. There was an almost Dickensian atmosphere about the place. We regularly played croquet on the lawn at the Rectory.

One of our immediate neighbours in Carlow was the choirmaster at the Protestant (Church of Ireland) church and he persuaded me to join the choir. I therefore became a chorister again and remained one - although in a fair variety of churches - for most of my working life. While in Carlow, I also joined the local choral society whose director was the organist from the Roman Catholic cathedral. I was the only Protestant in the choral society, so - although the Republic was a much more tolerant place than Northern Ireland - there were limits even there.

Herb, the choirmaster of the Church of Ireland church, was English but had lived in Castleblayney near the Northern Ireland border before coming to Carlow. He was on the select vestry (church council) of the church there and, one evening, raised the matter of the method of taking up the collection in the church. The procedure there was that the various chaps taking up the collection would deal with the pews allocated to them and then wander up to the altar as and when they had finished taking up their bit of the collection. Being a

tidy-minded individual, Herb suggested that it might be better if all those taking up the collection were to gather at the back of the church until they were all ready and then make an orderly procession up to the altar. One of the other members of the select vestry interjected,
"We want none of your damned Popery here!"

There was a nice story circulating in Irish musical circles about an amateur production of Rigoletto. The leading lady was the "classical" type of soprano, as broad as she was long and weighing about twenty stones. Rigoletto was rather diminutive. When it came to the scene near the end where she had been stabbed and was in the sack and had to be dragged across the stage, the operation was obviously providing Rigoletto with some difficulty. A wag from the back of the hall shouted,
"It's no good, you'll have to make two trips!"

The house where we were living was one of a group of eight that had been built very recently. We were soon on the best of terms with all our neighbours with the exception of the people opposite, an Irish doctor with an English wife, both Catholics. She came from Nottingham. Unlike Betty, who immediately got on extremely well with all our Irish neighbours, the lady from Nottingham did not settle in very well. Betty was very worried that she was not friendly with the lady because, after all, she came from Nottingham and, relatively speaking, had been almost a neighbour in England. Full of Christian spirit as the first Christmas approached, we thought that we ought to do something about it so, as there was no midnight service in the Protestant church, I went across the road to offer to baby-sit on Christmas Eve so that they could go to the midnight mass. What I did not know, unfortunately, was that many people had arrived drunk at the midnight mass on the previous year and, in consequence, the bishop had decided that there would not be any midnight mass on this particular year. The lady assumed that I knew all about this and took my offer as being a carefully calculated insult by a "Black Protestant" from the North. Sometimes you just cannot win!

The combination of Irish blarney, and the fact that the blight of television had not yet arrived, meant that conversation was a delight. One of the more pleasant special events was the custom of having Canasta parties at home. They usually consisted of four couples and for convenience we simply divided so that the men played against the women. The game really only formed a backdrop to about three conversations going on simultaneously and gently circulating round the table. In order to keep pace with what was going on, it was necessary to keep track of all three conversations - in addition to watching the game. An additional spice was added by the fact that one of our number regularly, if gently, cheated and had to be watched. Quite an educational exercise!

Some of us used to play badminton in the evening until about 10 pm, when we went to a particular pub for a drink. Closing time was also at 10 pm so we used to knock at the side door and then join a select group of late drinkers in the kitchen. A few years later, the licensing laws were relaxed to meet the needs of the tourist trade and the new closing time was 11.30 pm. These new hours, however, were rigidly enforced. By that time, we had left Carlow but I was back on a visit so I popped into the pub for a drink and asked the landlord what he thought of the new licensing laws.

"Disgusting," he said.

On my enquiring further, he explained.

"I know my customers very well," he said, "in the old days, I would put any of them who might not be able to hold their drink out through the door at 10 o'clock. Those who could drink like gentlemen could retire to the kitchen. Now, every Tom, Dick and Harry can drink until 11.30 and I have endless trouble when it comes to putting them out at 11.30."

There is a moral here somewhere that I suspect is not unconnected with some present difficulties in England!

Under the old laws, it was forbidden to sell drinks on Sunday except to bona fide travellers. The regulation that fixed whether you were a bone fide traveller or not was that you must be more than two and a half miles from home. As a consequence, some pubs outside the town were referred to as being "just a nice bona fide walk from town".

The Protestants had their own hall in the town, the Deighton Hall, where they held dances and so on. Whenever there was a dance, the townspeople would be joined by farmers and their families, who came from some quite considerable distance out in the country. There was a dance in October one year and because there had been some interference with motor cars in the town, Robbie - a friend of mine who was one of the organisers of the dance - asked the Civic Guards to put a Guard on duty to look after the cars parked near the hall. It was a rather cold and miserable night and presumably the Guard got bored because he walked along the parked cars and checked their licence discs. A high proportion of the discs had expired at the end of September (they had only been licensed for the quarter) and their owners received summonses for non-payment of tax. Robbie was not very popular!

As already mentioned, it was only necessary to cross one field at the back of the house to arrive at the River Barrow. At that time - and perhaps still - the Barrow did not suffer from much pollution and was very pleasant for swimming purposes so we did not have far to go for a swim on a summer evening. About a mile upstream, there was a small tributary that had a sandy bottom and was not too deep - ideal for teaching children to swim. About a mile downstream was the boat club, where it was possible to take out a rowing boat and spend a few hours on the river. Altogether very pleasant.

On one occasion, Betty and I were entertaining two engineering colleagues from a research institute in London. I mentioned the river and they asked me if there were any fish in it. I pointed out that there was only coarse fish in the river and that in order to fish for trout, it was necessary to go upstream about a mile to the tributary. Worse still, the nearest place where it was possible to fish for salmon was about seven miles away.

"You fish, don't you?" asked one of the visitors.

When I replied in the negative, he was quite horrified. I gathered that he had to travel a long way and pay quite a lot of money for some very indifferent coarse fishing.

One of our little group of houses was occupied by Abie, who was Jewish and - among other things - ran the local second-hand furniture store. He had been very helpful to us when we were furnishing our house. His father had come to Ireland from somewhere in Eastern Europe many years before and the old gentleman was still alive. He had one or two interesting peculiarities. As a devout Jew, he knew all the Psalms off by heart in Hebrew and liked to quote from them occasionally in conversation. However, when he wanted to quote from the Psalms, his method was to translate first into Yiddish and then from Yiddish into English. The resulting version was often an interesting contrast with that to be found in the Book of Common Prayer! Another of his peculiarities was that he used to wander into the second-hand furniture store, now being run by Abie, and offer advice to the customers. There were two difficulties about this; one was that he seemed to have forgotten whose side he should be on and the second was that he appeared to be completely unaware of the inflation that had taken place in the preceding few decades. His advice therefore tended to be something like,

"You can't pay five pounds for that, it's not worth more than a pound."

Fortunately, Abie was a very tolerant man.

At one point, there was great excitement among the Church of Ireland community. We were going to have a curate who came from South Africa and this curate would be taking the religious education in the little Church school. The children were agog on the day the curate was supposed to arrive, expecting an enormous Zulu or someone like that. Unfortunately, when Rexy arrived, he was very small, very white, rather pale and wore thick spectacles - quite a disappointment for the kids.

The arrangement under which Rexy had come to Carlow was that he would do some church duties and earn sufficient money to enable him to attend Trinity College, Dublin on several days of the week in order to study Hebrew. At a party in our house, Rexy met Abie and discovered, to his delight, that Abie knew a lot of Hebrew. After that, it was a very common sight to see this ill-associated pair together at social functions in Carlow. The large and florid, fun-loving Abie and the small and abstemious Rexy would sit together and earnestly discuss some of the finer points of Hebrew grammar.

At that time, back in the Fifties, getting to university was nothing like so commonplace as it is nowadays. The son of one local family of fairly modest means had succeeded in getting to Trinity

College, Dublin by winning a scholarship. Even so, it must have meant a substantial sacrifice for the family. Mother, however, could not get over the fact that her son was actually at the university and when the lad was at home, he only had to make some remark such as "I think it's going to rain" to be rewarded by his mother with some remark such as,

"Ah, the trained mind" or "The benefits of an expensive education".

This latter quotation has now, in my family, become the standard response to any particularly platitudinous remark.

About this time, there was a murder in a small place about fifty miles from Dublin. That, at least, is true and it would be nice to think that the rest of the story is. At any rate, it was said that after he had done the deed, the murderer was immediately overcome by remorse and went to the Civic Guard station where he found the solitary Civic Guard (it was only a small place) sitting on his chair with his feet up on the desk and his collar undone.

"Guard," said the man, "I want you to take me in, I've just strangled a woman."

The Guard reluctantly removed his feet from the desk and straightened himself up.

"Ah now," he said, "a joke's a joke and all that but I think that is in very poor taste altogether."

After a fairly long argument, the man eventually convinced the Guard that he really had strangled a woman.

"Well," said the Guard, "a right mess you've got me into. I was supposed to have been retiring tomorrow!"

My job involved quite extensive travel around Ireland and on one occasion, I found myself in County Clare quite near to the famous Cliffs of Moher. As I had never seen them, I went to have a look. There are, I am told, people who are prepared to stand on the edge of the cliff and look down. I crawled towards the edge on my tummy and peered over. The cliff drops sheer for 1200 feet to the Atlantic with the Atlantic waves breaking furiously at the bottom.

Having returned to the road where I had parked the car, I was buttonholed by a chap who really was a character - even by Irish standards. He cornered me, as I gather he did everyone else who came that way, and complained bitterly that when he was in the LDF (the Irish wartime equivalent of the Home Guard), the unit commander would take no notice of his suggestion that they should erect a concrete machine gun post at that point. He had, he said, pointed out that if the Germans had landed, the road could have been well covered by the machine gun post at the point and they would never have got up the road. In view of the fact that the only thing the Germans could have done - if they had gone up the road - would have been to jump over the cliff, I thought the unit commander might just possibly have had a point.

It was while we were living in Carlow that our third son arrived and this was associated with a rather odd event. Betty's younger sister, who lived in Kent, was pregnant at the same time as Betty. In due course, Betty started labour pains one evening and I took her into the nursing home. Everything seemed to be going extremely well and the pains were becoming very frequent until, about 4 o'clock in the morning, the pains suddenly stopped. There was no further activity whatsoever and about midday, I had to bring Betty home again. Later on, we discovered that Betty's sister had given birth to her baby at precisely the time when Betty's labour pains stopped. Betty's baby did not arrive for another fortnight.

Socially and in many other ways, life in Carlow was ideal but things are seldom perfect in life and there was a snag. I had originally come to Ireland to develop and then put on the market a device for burning peat (the only fuel available in Ireland at the time) in a special burner that could be attached to boilers and similar equipment. The device was already available as an experimental model produced by an Irish research establishment. Unfortunately, between the time when I had accepted the job and when I actually arrived in Ireland, the government authorities had decided that the Carlow company could not have an exclusive licence because the device had been produced by a government-financed research establishment. It was available to anyone who cared to manufacture it. As a result, quite a few companies decided to put it on the market in the prototype form - as available at the research establishment - and the Carlow company, rather against their better judgement, had to follow suit in order to get some share of the market. The inevitable result was that a completely undeveloped device was installed in many locations and gave the trouble that might be expected. The result was a financial disaster, although I did manage to redesign the equipment for some special installations on a sounder engineering basis and, as far as I know, these proved quite successful.

One establishment where I installed one of these special burners on the central heating boiler was a Roman Catholic college in Killarney. After the necessary work had been carried out, the bursar came down to have a look so I told him how it worked and showed him the flame. He said it looked very nice but how did he know that it would heat the college? That was rather difficult to demonstrate, it was the middle of summer - it is not a good idea to mess about with central heating equipment in winter - and the college was fairly warm anyhow. When did I think the college would be at its coldest? I shrugged my shoulders and rather unwillingly volunteered the information that it should be at its coldest at about 2 o'clock in the morning.

"Right oh," he said, "I'll get the nuns to give you a bit of dinner and then you can come up to my room and we'll have a wee drink together and try it out at 2 o'clock in the morning."

The nuns gave me a very good dinner and then I went up to Father Long's room where he offered me a drink of whiskey and we had a chat. It soon transpired that we had several interests in common, including an interest in languages. His mother tongue was Irish, and English was therefore his second language. He had studied at the Irish college in Rome and, in fact, taught Latin and Greek at the college. Apart from that, to my knowledge, he could speak Italian, French and German. Time sped on fairly rapidly and suddenly about ten minutes past midnight, he announced that it was time he said the office of the day and since - as a good Protestant - I would not wish to listen to that, he offered me a Bible to read while he was saying it.

"Just a minute," I objected, "you should have said the office of the day before midnight."

"Ah," he replied, "Here in Killarney we are twenty-five minutes west of Greenwich. The Lord goes by the sun, not by Greenwich Mean Time!"

At 2 o'clock in the morning, I went down and started up the boiler; he dashed round to the remote corners of the college to ensure that the system really was effective.

Father Long, later Archdeacon Long, remained a good friend through all the years. Although we only met every ten years or so, we exchanged correspondence once or twice every year and when we did meet, it really was a meeting of old friends. He died a few years ago.

At one point, there was interest from an Irish government department in installing a particular type of boiler and fitting it with our combustion equipment. Unfortunately, there was no boiler of this type in the Republic of Ireland and the nearest one was in Ballymena in Northern Ireland. Accordingly, four of us set off by car early one morning to travel to Ballymena to see this boiler. In addition to myself, the party consisted of another Ulsterman from the Belfast area, an extremely respectable civil servant from Dublin and Frank, whom we met earlier at the distillery and who was now with Bord na Mona (the Irish Peat Board). Frank had strongly nationalist opinions. Once we had crossed the border and were travelling through the fairly rich farmlands of North Down, the two of us from that area started to make appropriate comments on the fact that it was obvious we were in the North from the neatness of the hedgerows.

"Stunted in their growth through British oppression," said Frank from the back seat.

As we had made a very early start that morning, we had completed our inspection of the boiler before lunchtime and the four of us headed into the centre of Ballymena and found a hotel that had been recommended to us for lunch. As we passed through the front door, we were met by a gentleman who, as we thought, enquired,

"Four?"

We agreed that that was indeed a good description of our party and he directed us to a door on the right. On entering the door, we found a large party in progress and drinks were promptly thrust into our hands. Needless to say, we did not object. One chap came up to us and enquired where we came from and, when we told him, he expressed surprise that there were representatives there from the South. This seemed a good point at which to enquire what was going on and we discovered that it was a meeting of the *Ford* agents in Northern

Ireland. We explained our mistake and were going to make ourselves scarce but everyone thought it was a wonderful joke and we could only escape after having one or two more drinks. Drink/driving laws still lay far in the future!

Despite any differences in political opinions, Frank and I became quite good friends. He told me one rather amusing tale about a friend of his of extreme nationalist views who had decided to investigate the origins of the tune "The Londonderry Air". The origins were, and remain, rather a mystery and anyone with nationalist views would find it peculiar that this typically Irish tune should have the name "Londonderry Air" rather than "Derry Air" – Derry being the Irish name for the town. As a result of his researches, the extreme nationalist discovered that in the previous century, the Lord Londonderry of the time - who was quite an admirer of Irish traditional music - had organised a meeting of Irish itinerant musicians. He apparently did them proud and one of the fiddlers present composed a tune in honour of the occasion and named it "The Londonderry Air". The Londonderry family have the reputation (in Ireland at least) of being anything but pro-nationalist and so, as my friend Frank wryly remarked,

"The results of this piece of research are unlikely to be published".

Another story, which I think came from Frank, concerned a farmer with a cow to sell at the cattle market. The dealer examined the cow and said:

"I'll give you 95 pounds for her less 15 per cent."

The farmer found the necessary arithmetic a bit complicated and said that he would think about it. He adjourned to the pub and ordered a drink but the calculation was still beyond him so he said to the barmaid:

"If I was to give you 95 pounds less 15 per cent, what would you take off?"

"Everything," said the barmaid, "except my ear-rings and they're awkward."

There was one rather big job of which I was really quite proud. We were asked to obtain boilers for a fairly large company and equip them with combustion equipment of our own manufacture. For reasons of their own, the company insisted that the boilers should be of Danish manufacture. From the information supplied by the company, I produced specifications for the boilers and drew the combustion equipment in sufficient detail to show the connections necessary between the combustion equipment and the boiler, although some details would have to be discussed with the boiler designer before the design of either could be finalised. I carried out my design work in accordance with the appropriate British Standard Specifications. My specifications for the boilers called up the same standards. There were two Danish companies that were suitable as suppliers and they both had head offices in Copenhagen. It was agreed that I should travel to Copenhagen and clear up all the technical points with both companies so that they could quote appropriately.

On arrival at Copenhagen airport fairly late in the evening, I was - to my surprise - met by a representative of one of the companies. He drove me to my hotel and once I had checked in and unpacked my luggage in my room, he took me off to dinner. He then took me to a nightclub and in due course drew my attention to some of the charming ladies present and indicated that I could have my pick. Since - apart from any moral considerations - I had not the slightest intention of being bribed in this way, I began to talk very pointedly about my wife and children and the benefits of family life in general. He then took me back to my hotel. By the following evening, he had obviously decided that his initial démarche had left something to be desired and he rang me up full of apologies and asked whether I could go out to dinner with another member of his staff. I demurred for that evening but agreed to go out the following evening. It was a young couple who came to collect me on that evening and the company had obviously been dredged to find the most respectable couple in the place. They were quite incredibly dull and took me out to dinner in a very stuffily respectable restaurant. As if that was not bad enough, they insisted on taking me back to their home so that I could admire their newly-born baby. "What I do for Ireland," I thought to myself.

In the meantime, I had had serious business meetings with both this company and the other company.

During the meeting with the second company, I was seated on one side of a table and a Danish engineer from the boiler company was on the other side. I carefully unpacked what was necessary from my briefcase. After some initial chat, I rested my right hand on the little pile of British Standard Specification documents there and pontifically enquired whether my opposite number was familiar with these British Standards. He agreed that he was and we started work. He was an excellent and very competent man and we got through the business very rapidly indeed. I agreed that we had settled everything and started to repack my briefcase. When I opened it, I found to my horror that my copies of the British standards were still in my briefcase - I had been pontifically laying my hand on his copies of the specifications! Feeling rather a fool, I admitted what I had done. He was greatly amused and we had a good laugh together.

When the quotations finally arrived, there was not much difference between them and I am offering no prizes for guessing who got the job.

The installation - both the Danish and Irish parts - was, to the best of my knowledge, successfully operated for many years.

Despite my efforts, the original job for which I had come to Ireland folded up and although the company were extremely generous and continued to employ me in various capacities, I did not feel that I was earning my keep and, quite certainly, there were no career prospects. On the other hand, the experience of working in a small company where the responsible engineer had to deal with all aspects of the work from the original quotation to the final invoicing, including design work and supervision of manufacture, proved invaluable later in life. Be that as it may, the summer of 1956 saw us leaving the Republic of Ireland and returning to Derby.

Rolls-Royce, Derby - second stint

So the summer of 1956 found us returning to England. We were, by now, a family of five with three boys, James, Ian and Andrew. Like many other English people who have gone to live in Ireland, Betty had been initially nonplussed by the differences but she eventually became more Irish than the Irish and was most reluctant to return to England - even though she was returning to her home ground. The difficulty was compounded because the only house I could lay my hands on rapidly was not nearly as nice as the one in Carlow and it proved quite impossible to find a house at all in her native village. However, you can't win 'em all. Rationing, at least, was a thing of the past in England.

The job I had acquired involved responsibility for "turbine blade cooling", a rather specialised field that was then coming into prominence. The job involved close liaison work with the Rolls-Royce branch factory at Barnoldswick, with sub-contractors in Glasgow and Sheffield and with the National Gas Turbine Establishment at Pyestock, near Farnborough.

One of the engineers at Barnoldswick was redecorating the living room in the stone cottage he had acquired. In order to paper the living room, he needed to smooth the stone wall by pointing, plastering and so on. However, there was one very large stone near ground level, which protruded a long way into the room. He tried to cut through it but soon found that it was far too hard and decided that there was nothing else for it - the stone would have to come out. Chipping away the pointing, he eventually loosened the stone and fetched it out, leaving a very large black hole. Being of a curious disposition, he thought he had better investigate. Bending down and pushing his arm into the hole, he found it was much larger than he had thought and began to have romantic visions of buried treasure or the like. Eventually, when his arm had disappeared practically up to the shoulder, he found he was holding a piece of cloth. He pulled it out and saw that he was holding a piece of green baize - perhaps what the treasure was wrapped in! He persevered and eventually his fingers touched a piece of wire. He pulled it and it gave out a musical note.

He had pulled the back off the piano in the house next door!

My job involved convincing staff in other departments that they should do various things to help the blade cooling business along. Fortunately, my stay in the south of Ireland must have softened my natural taciturn northern nature and, on one occasion, one of the designers - whom I was trying to persuade to do something my way - commented that I must have kissed the Blarney stone. One of his colleagues then butted in,
"Kissed it! He swallowed the b...."

Be that as it may, I was in the design office on one occasion and, rather than making an immediate effort to cajole one of the designers into doing a rather awkward job, I mentioned the matter of the wallabies as a conversational opening. At that time, it had just been revealed that there were wallabies living wild in north Derbyshire. They had escaped or been released from a private zoo and, much to everyone's surprise, had found themselves a niche in the Peak District ecology. One of the designers enquired what a wallaby was and I informed him that it was like a small kangaroo.

"Thank God," interjected another designer.

We enquired the reason for this sudden gratitude to the Almighty and he informed us that during the previous winter, he had been driving back from Barnoldswick with three colleagues who were dozing gently – perhaps because they had stopped somewhere for a few drinks. Suddenly, an animal became visible in the headlights as it dashed across the road.

"My God," our friend had shouted, "A kangaroo!"

This wakened the other three who decided that he was in no fit state to drive and insisted that someone else drove.

"I've never heard the end of it since," he said.

On one visit to the sub-contracting company in Glasgow, there was a one-day's delay in getting on with the business in hand and the host company very generously sent my boss and I on a jaunt round the countryside in a comfortable car. One of the places we visited was Largs on the Firth of Clyde and the driver took us to see the cemetery - which gives me the opportunity for another graveyard story. The Largs cemetery is on a fairly flat piece of ground about half way up a mountain behind the town. There is a quite incredibly beautiful view from there, over the Firth of Clyde, to the islands (Arran, Ailsa Craig, and so on) and to the Highlands to the north. We were told that a favourite walk for older people from Largs was to go up to the cemetery, where they could potter around the graves of their friends and relatives, do little odd jobs and admire the view. The story, however, concerns three old gentlemen who were up there on one glorious Sunday afternoon.

"I was 72 last Thursday," said the first old gentleman. "I was a sailor, you know."

"Don't we just," said the other two.

"Whether you like it or not," retorted the first old gentleman, "I've been round the world at least three times - not straight round every time, ye ken, but all in all, I must have been round about three times. I've seen it all, Rio de Janeiro, Naples, Hong Kong - and very beautiful these places are - but I don't think I've seen anything more beautiful than this ..... and when my time comes, they can lay me down over there with my old friend John McNally and I won't complain."

"Well," said the second old gentleman, "You were very lucky. I was 84 a couple of months back but I've been a carpenter in Largs all my life. I've hardly been out of the place, apart from the odd trip up to Glasgow for the football, but I'm quite prepared to believe that you're right. There can't be many places in the world more beautiful than this and when my time comes, if they'll just lay me down over there with my old friend Harry McNab, I won't complain either."

"Och aye," said the third old gentleman, "I was 96 last November and I agree with both of you - it's very nice up here. In fact, I wouldn't mind lying down here with Jeannie McPherson."

"But she's no' dead yet!"

"Neither am I, neither am I."

Scotland must be one of the few places where I made a really determined effort to combine business and pleasure. A business visit was due to Glasgow and I fixed it for a Friday because there was a rugby international between Scotland and Ireland the following day at Murrayfield. The people at Glasgow were very co-operative and arranged to put me up in the company's guesthouse on the Friday night and they also managed to arrange some stand tickets for the match (I did pay for my own ticket!). "The best laid plans of mice and men..." Not only was the match played in a blizzard but the wind was blowing directly into the stand where we were sitting. We were rapidly covered in snow and all we saw of the match was the occasional lineout immediately in front of the stand and a few other bits of play in a similar vicinity. The only bright spot in the afternoon was that I was informed later that Ireland had won - from where we were, we had no idea!

The other main sub-contract company with which I was concerned was in Sheffield. They had a designer - an old chap (at least that's how he seemed to me then!) who designed forging dyes and could quite accurately be described as an artist in white-hot metal. He took me to lunch one day and I discovered, somewhat to my surprise, that he played double bass in the City of Sheffield Symphony Orchestra. We discussed musical subjects and I was silly enough to express the opinion that the more modern composers like Shostakovitch could not hold a candle to their predecessors - Bach, Mozart, Beethoven, etc. He looked at me very severely.

"Young man," said he in his strong Yorkshire accent, "Young man; if tha'd heard Shostakovitch as often as tha'd heard Bach, tha'd think he was aw'reet."

I felt about two inches tall, which served me right!

Business also took me frequently to the National Gas Turbine Establishment in Pyestock. I had acquired a rather ancient, pre-war Vauxhall car and was most anxious to drive it down to Pyestock on a business trip. This was in the days before motorways and it was quite out of the question to get there and back in one day - at least one overnight stop was necessary if any serious business was to be done. A glance at the map had suggested that Marlow-on-Thames would be a suitable place to break my journey and, borrowing an AA book (considerably out of date) from a colleague, I noted that there was a pub called the "Compleat Angler" at Marlow - bed and breakfast twelve shillings and sixpence. I asked the office secretary to book me a room. In due course, I arrived in Marlow and asked a passer-by for directions to the "Compleat Angler". He looked rather quizzically at my battered old car but informed me that it was just over the bridge on the left. I drove over the bridge and turned into a driveway on the left, which led through magnificent lawns by the river and up to a rather imposing-looking hotel. I parked in front of the hotel and went up to the reception.

"I believe you have a room for me, Stewart is the name."

"Oh yes, Mr Stewart of Rolls-Royce."

A flunkey appeared at my elbow.

"Shall I take your car round for you, sir?"

Quick think.

"Oh no, thank you, I prefer to drive it myself."

So I drove my battered old Vauxhall round the back, parked it between a Bentley and a Rolls and went into the hotel. My room was magnificent, en-suite and with a beautiful view down to the river. En-suite rooms were still a rarity in those days. However, I had a bath and went down to the bar where I ordered a beer. The price of the beer confirmed my worst suspicions and a glance at the menu did nothing to help. Having drunk my beer, I went up the town and had some fish and chips wrapped up in newspaper as my dinner in the hope that that would keep my expense account down to some remotely reasonable level. Relating the matter at the National Gas Turbine Establishment the next day, one of my listeners informed me that the Aga Khan had been refused service in the restaurant at the "Compleat Angler" the week before because he was not wearing a tie!

As part of my job, I had to make arrangements to have special instrumentation produced in the electrical laboratory. There was a key man in the laboratory who had been called up for military service in the Air Force. As he was a key man, the firm moved heaven and earth to try and prevent his departure. These negotiations went on for months and months. Eventually, the firm lost and we said goodbye to him with due ceremony one Friday and wished him the best of luck. A few days later, I was in the electrical laboratory and there was our friend!

"What are you doing here?" I asked.

He had been failed, for flat feet, at the medical!

Soon after I arrived back in England, I joined the choir in the local church. It was a "men and boys" choir in which most of the men had sung together for a long time and were beginning to die off. The organist did not seem to realise that the ranks were thinning a bit and carried on with the same repertoire. He also became rather deaf before he too died. He did his best to disguise his deafness by making detailed criticisms and suggestions to the choir. One evening, he stopped the rehearsal of a hymn and requested the tenors to sing more quietly. Ernie, one of the tenors with a very strong sense of humour, whispered to his colleagues,

"Let's mime", so we mimed the next verse.

The organist stopped again, looked pleased and said,

"That's much better, tenors."

A very patriotic man, the organist never used the Haydn tune (which is also the tune of the German national anthem) for the hymn "Glorious things of thee are spoken". One evening towards the end of the Fifties, however, this hymn was on the agenda and, as we stood up and the introduction to the hymn was being played, it was Haydn's tune which emerged from the organ. Ernie whispered,

"I think the war must be over".

During that time, I also joined the company's amateur operatic society. I have always had a strong suspicion that performances by amateur operatic societies are intended less for the enjoyment of the audience than for the amusement of those taking part. The amusement and enjoyment, of course, mainly applies to the lower ranks. As far as some of the "better" singers are concerned, it is an extremely serious business and not getting a part they had set their hearts on can ruin their lives for months on end.

The first show in which I participated was "The Student Prince". There were a couple of amusing incidents. One evening - at the point where the prince was supposed to come on stage in great excitement and say to Dr Engel "Oh Doctor, Doctor, it's wonderful, there's students all over the place and they're hanging lanterns all along the riverside" - he unfortunately got things mixed up and said, "Oh Doctor, Doctor, it's wonderful, there's lanterns all over the place and they're hanging students all along the riverside".

There was a non-singing part in the show - the old waiter at the inn at Heidelberg. As it was a completely non-singing part, none of the operatic society people wanted to have anything to do with it and we roped someone in from the company's dramatic society. He had one great moment in the second act when the prince had become king and the old waiter appeared, travel-worn and weary, having walked all the way from Heidelberg. He told the king a very sad tale about how the old inn had gone down in the world, Kathy (the erstwhile girl-friend) was always crying and so on. As the final pièce de résistance, he declared, "Ah, your majesty, I almost forgot. I've brought you a rose, a rose from the garden in Heidelberg." He then thrust his hand in his bosom and produced the rose, handed it to the prince and went off-stage. This went fine until the Thursday night when he thrust his hand in his bosom and a look of startled horror appeared on his face. He really had forgotten the rose! This complicated matters for the leading tenor, who was supposed to sing to the rose. However, he carried on without it – with the spotlight on him. Suddenly a hand appeared from out of the shadows and thrust the rose into the singer's hand! The effect on the audience can be imagined.

Power plant engineer, preliminary design office

After a few years on the blade-cooling job, there was a re-organisation in the firm and I became a power plant engineer in the preliminary design office with the main task of maintaining liaison with European aircraft manufacturers. This involved considerable travel in Western Europe, mainly France, Germany and Italy. It was during this period that I made the acquaintanceship of the Guidet family in Paris - an acquaintanceship that gradually deepened into a friendship between the families and lasted through the years. It was also during this period that our daughter was born and that completed our family - three boys, James, Ian and Andrew and a girl, Siobhan.

A frequent companion on my travels abroad was a designer called Maurice Taylor. We had various little private amusements of our own. It was about this time that the French currency was re-organised and, somehow or other, one of us was left with an old five franc piece - a fairly large coin which was just about worthless. Whenever we arrived in Paris, the one who did not have this coin had the duty of finding a trolley for the luggage and pushing the luggage through the airport. At the point where we parted with the trolley, the possessor of the five franc coin would solemnly hand it over to the other as a tip and be thanked by the other in an ingratiating manner, thus convincing the foreigners that the British really were mad.

One of the places that we regularly visited was Friedrichshafen on Lake Constance. Although

Friedrichshafen is in Germany, the most convenient way of getting there at that time was to fly to Zurich, take the train from there to Romanshorn in Switzerland and then take the ferry across the lake to Friedrichshafen. One day, Maurice and I were going across on the ferry and struck up a conversation with three Dutch civil servants, who told us that they were going to Friedrichshafen to attend the IBO Messe, a very important agricultural fair. In Friedrichshafen, we met Reggie Longinotto who was one of the company's senior representatives in Europe and normally resident in Geneva. Reggie had Italian ancestry but it was quite a long way back. Going into the bar in our hotel in the evening, before dinner, we found the three Dutchmen there so, of course, I introduced Reggie. One of the Dutchmen said that Longinotto was a peculiar name for an Englishman.

"Oh no," said Reggie, "I'm not with these chaps, I've come for the IBO Messe. Surely you have heard of the Longinotto spaghetti machinery."

The Dutchmen confessed that they had not.

"I am not surprised," continued Reggie, "Do you know, the Dutch eat less spaghetti per head of population than anyone else in Europe."

They looked suitably contrite and worried. One of them asked what was new in the spaghetti machinery industry. Reggie started by mentioning the fact (actually true) that some American manufacturer was attempting to put square spaghetti on the market. It was said to be easier to grasp with the fork.

"Such an idea," said Reggie, "Square spaghetti! No imagination! No fantasy! But we have developed a machine that produces spaghetti with a shamrock-shaped cross-section. Can you imagine anything that would sell better in America that a combination of something Irish and something Italian?"

At that point, we left them and went to dinner. I met two of the Dutchmen at breakfast the following morning. They were still worrying about the shamrock-shaped spaghetti.

On a visit to Turin in Italy, several of us were entertained by an Italian contact in a rather nice restaurant. Over one of the tables, there was a plaque to commemorate the fact that Emile Cavour used to lunch there when the Piedmont parliament was in session. One of my English colleagues said,

"Who was Emile Cavour?"

The Italian looked rather hurt that my colleague had not heard of one of their great men and, I must say, my colleague might have been better to keep a tactful silence. Fortunately, my old-fashioned grammar school education stood me in good stead and I was able to intervene,

"There were three people mainly responsible for the unification of Italy. Garibaldi did the fighting, Mazzini did the shouting and Cavour was the smoothie who fixed it."

Our host smiled.

"We do not usually put it that way but your summary is quite accurate."

(Blessings on Bunny Haire and Miss Tipping who shoved the history down our throats!)

Maurice and I visited a research institute in Spain at one point. The Spaniards had obviously heard that the British could not survive without tea so they had given instructions to the canteen to produce tea in mid-morning. The ladies in the canteen had dutifully made the tea and then set out on foot to deliver it. Unfortunately, the canteen was about a mile from where we were having the meeting so that the tea was stone cold when they arrived. Our Spanish hosts were well aware that tea should not be cold so they made rapid arrangements with the laboratory next door. The milk was poured into the teapot and the whole lot was boiled up thoroughly. It was then poured out for our delectation. "Delicious", we said and forced it down.

The customer is always right.

It was about this time that Rolls-Royce decided, for some reason or other, to form a company in France. It had exactly the same premises, facilities and staff as the previous Rolls-Royce branch but the manager of the branch was made a director of Rolls-Royce (France). He pointed out that as a director of the company, he should be provided with a Rolls-Royce like the other directors. There were initial objections that he was not a 'real' director but he persisted and, eventually, he obtained his Rolls-Royce – albeit a second-hand one. Shortly after this, two of the company's senior engineers, Geoffrey Wilde and Tom Kerry, had to pay a visit to Paris. Normally, they would have been told which hotel they were staying in or, if they were lucky, the 'lad' from the office would be sent out with the Ford to collect them. On this occasion, however, they were met by the boss in person. When they got to the car park, it was obvious why. There was the Rolls-Royce. Tom whispered to Geoffrey,

"We'll both get in the back."

This they did, leaned back in the luxurious upholstery and Tom said,

"Hotel Gallia, Driver."

It was also about this time that three of us paid a visit to an aircraft company in Sweden. It was in the latter half of August and, had we but known, it was during the "crayfish season". The Swedes were only allowed to fish for crayfish during a three-week period and they had their own customs associated with the crayfish – as we were to learn. There was a message for us at the hotel to say that, because of the large amount of work to be got through, a car would call for us earlier than usual the next morning. "The customer is always right" so we breakfasted early and the car duly arrived at an early hour.

At the beginning of the meeting, the chairman announced that in view of the large amount of work, we would not be going out to lunch but that sandwiches and so on would be brought to the conference room so that we could work straight through. Again, the customer is

always right, so we agreed. We worked very hard all morning, and while we were eating our sandwiches – it was then about 2 pm - the chairman stated,
"Well, we seem to have finished."
He was, indeed, right. We had crossed every possible "t" and dotted every possible "i". He then continued,
"It really is a beautiful day and we thought you might like to go for a swim. Herr Dilner will take you. "

Herr Dilner took us to a beautiful lake that was not too far away and had us swimming back and forth across it. (We later found out that although he was not actually an Olympic swimmer, he had got as far as the Swedish trials for the Olympics.) He then drove us back to the hotel where we were handed over to another of the Swedish engineers, who had another surprise in store for us.

This chap stated that he was very anxious to show us the Douglas castle, which was not too far away. As he drove us there, he told us that Douglas, although a Scotsman, was in fact a successful Swedish general who had built himself a castle on top of a fair-sized hill and he thought we ought to see it. When we got to the castle, we climbed up all the stairs to the top of it and then up a ladder to get to the top of the tower, where there was a magnificent view of the countryside and a distinctly fresh breeze. This stimulated an already present appetite to something approaching a ravenous hunger. We then climbed back down the tower and down the stairs of the castle and back into the car.

He drove us back to the hotel, where we observed that the other Swedish engineers were arriving by taxi.

We might not know much about the crayfish season in Sweden but we did know that, even at that time, the Swedish drink-driving laws were extremely severe and that the arrival of the other engineers by taxi could mean only one thing. We adjourned to the bar where, in my honour, large quantities of Stewarts' whisky were produced. By this time, we were furiously grabbing any of the crisps, pretzels, etc which were available in the bar. The Stewarts' whisky session seemed to last quite a while but eventually we migrated to the dining room.

The meal started with crayfish and we were introduced to the custom associated with eating the crayfish. The idea is that for every crayfish claw that you eat, you drink a schnapps. In case you should feel thirsty, there was also beer available. There is about as much meat in a crayfish claw as would cover the fingernail of your little finger!

After we had dispensed with the crayfish, we were rather hoping for something substantial like a steak would appear. Not a bit of it. The next course was a very light soufflé accompanied by Madeira wine. After that, we had coffee and brandy and then we were informed that it was really time to have a go at the schnapps.

The memory starts to get a little bit dim from then onwards but I remember that, at one point, the Swedes started to sing folk songs to the schnapps – apparently a traditional custom. We were then invited to sing an English folk song to the schnapps and we did our best with "On Ilkley Moor bah't 'at".

Eventually, the dining room was deserted apart from our party and the Swedes announced, with some disappointment, that they really would have to go. On the way towards the door of the hotel, the dim idea penetrated my brain that "these characters thought that we would finish under the table". So I said,

"I have a bottle of whisky in my room, who is coming up for a drink?"

One of my English colleagues volunteered, together with Herr Dilner and another engineer who was Finnish (traditionally, I believe, a hard-drinking race).

In fact, I had been bragging. It was only half a bottle of whisky. At some point, my colleague and the Finn disappeared and Herr Dilner and I finished the whisky. I said good-bye to him at the door of my room. Although I knew that politeness demanded that I should accompany him to the door of the hotel, I was sufficiently compos mentis to realise that although I might get down the stairs – very rapidly – my chances of getting back up were very slim.

My hangover the next day is something I shall never forget.

One of the features of Derby at that time was the large number of Poles living in the town. One of them was a neighbour of ours and when I asked him why they had settled in Derby, he told me that those who did not wish to return to Poland after the war were given the choice between a large number of towns and cities in Great Britain where they might settle. Most of the names meant absolutely nothing to them but many Poles were rather fanatical about horses and had heard of the Derby. So they opted for Derby unaware of the fact that it had nothing whatsoever to do with the horseracing.

This friend of mine had fled from the eastern part of Poland in the early days of the war when the Russians moved in. He had with him the name and address of his much older sister who was married and lived in the United States. Because this piece of paper was much too dangerous to carry with him, he destroyed it - thinking that he would be able to remember the address. In due course, he found his way to North Africa where he joined the Polish army and was involved in the various campaigns in which the Polish army took part. After the war, he settled in England and married an English girl but by that time he had forgotten the address and could only remember his sister's married name and the fact that she lived in a certain town. He thought that it would be quite hopeless to try and trace her with so little information. However, one day in the late Fifties, he was walking past the general post office in Derby and thought that he should at least go in and enquire whether there was a place of the appropriate name in the United States. The counter clerk was most helpful and insisted that he go into the postmaster's office to see if he could get any help there. The postmaster produced the enormous international post office guide and, after some study, announced that there were six possible places in the United States. He carefully wrote out the details of these six places and gave the list to my friend - John was his name. (He had long since given up trying to use his Polish name, no-one could spell it.) The head postmaster suggested that he wrote to the postmaster at each of these places in turn enclosing a short letter to his sister and, who knows, he might just be lucky.

John picked one with a pin and sent off the letter. He had a reply back from his sister about two weeks later!

John and his family were soon invited to the States for a holiday and he was again lucky. At a dinner party, he was introduced to a local entrepreneur who asked him what he did. He mentioned that he had worked for some years in a factory specialising in a particular field. It then appeared that there was a proposal to set up a factory for this purpose in that particular area of the United States. He was asked if he could set up the factory and he agreed to undertake the job. While working his notice in Derby, he carefully noted the addresses of the firms who had manufactured the production machinery. Armed with this information, he was very successful and not only set up the factory but managed it for many years before retiring.

It would be wrong to give the impression that contacts with Ireland ceased altogether on the move to Derby. Apart from regular holidays that the family spent with my mother and father in Bangor,

I did have some business trips to Shorts Aircraft in Belfast and we even managed to pull in the odd visit to our friends in Carlow.

On one visit to Belfast, I was clear of the overnight boat rather earlier than expected so, having to wait for the hire car to arrive, I went into the little snack bar for a cup of tea. At this juncture, it is appropriate to mention that the Stewart family were famous for their liking of weak tea; in fact our version of tea was referred to by some friends as "water bewitched and tea begrudged". And I was the weakest of the weak tea brigade. Standing in the queue at the snack bar, I observed what the man in front of me had obtained in the way of cup of tea so, when it was my turn, I said,

"Tea, please, very weak and very little milk."

The lady behind the counter produced an enormous jug of milk which she started to pour generously into the cup. I made her pour most of the milk back into the jug. She then started to pour out the tea from her teapot - it was just about black - so I permitted her to pour about two cubic centimetres of tea into my cup and then asked her to top it up with hot water. It looked just about right to me. I thanked her and asked how much I owed her. She folded her arms over her capacious bosom.

"That", she said in a strong Belfast accent, "will be nothing. I sell tea here, I'm not taking money for that stuff."

On a visit to the Republic, Betty and I went to the Gate Theatre in Dublin and, during the

interval, we got talking to a gentleman who was a fairly senior official with an important Irish company. Not long before, someone had been going through their pension fund records and suddenly exclaimed that Sean O'Sullivan (or some such name) was 105 years old. There was immediately great interest - just the topic for the house magazine - and the finder was sent down the country to interview Sean at his retirement address. The chap charged with the job eventually found the little cottage and there was an old man sitting smoking his pipe outside the front door – old, but he hardly looked 105.
"Are you Sean O'Sullivan?"
"No, I'm Seamus O'Sullivan. Sean O'Sullivan was me father, he's been dead this many years since."
"But someone has been drawing his pension!"
"Oh yes, that's me. He left it to me in his will."

In 1963, we had at last managed to buy somewhere to live in Breadsall, which was Betty's home village. We moved in at the end of the year and, almost immediately afterwards, I was sent to Germany on a course to brush up my German. It was quite an intensive course but we did have three afternoons a week free, as well as all day Sunday, so I managed to learn to ski - quite apart from improving my German. I might have guessed that I was not being sent on this course for fun because, very soon afterwards, it was made clear that the company wanted me to become their technical representative with the German Ministry of Defence and be stationed near the relevant government departments in Bonn and Koblenz. In consequence, we had scarcely got to Breadsall before we were on the move again.

Technical representative in Germany

In the summer of 1964, therefore, the Stewart family set off for Germany. James, the eldest boy, was on holiday in France with the Guidet family and he travelled direct from there to Germany. After a full month in France, his French had become quite fluent and he was very proud of it. Much good it did him, he had to start to learn German right away!

Although Betty and I had already found a house, on a visit specially arranged for that purpose, the company's arrangements for moving our furniture took rather a long time - about six weeks - and during that period, we lived in a rather nice hotel in Bad Godesberg. This hotel had a garden restaurant and, in the evening, it was possible to have dinner under the chestnut trees while listening to the orchestra. It was obviously a place where the children had to be on their best behaviour and, although we had introduced them to restaurant eating from an early age, we tended to be a bit jittery. (Let's be honest, I was jittery, Betty was quite unflappable).

One evening, my main course was a very nice fish and, when I was about to start it, one of the children acted in such a way that I thought something would be knocked over. Making a rapid movement to avoid the catastrophe, I somehow managed to engage the fish with my fork in a rather violent manner which led to its going into orbit and landing in the chestnut tree above my head - much to the amusement of the children and everyone else in the immediate vicinity. I have never lived it down!

In due course, we received notification that our furniture was ready for delivery and we could therefore move into our home in Niederbachem, a village near the left bank of the Rhine and several kilometres to the south of Bonn.

Before dealing with that matter, however, I should relate something about Rolls-Royce crates.

When Rolls-Royce engines were being despatched, they were enclosed in very carefully made and quite expensive crates - as indeed they should be, aircraft engines are expensive and quite sensitive pieces of machinery.

One of the perks of Rolls-Royce employees was that they could buy old crates very cheaply and many employees did so in order to build garden sheds and the like. The story goes that one chap bought a crate in the autumn of one year and it was duly delivered to his garden. The weather broke immediately and he did not, therefore, do anything about it until the following spring. On opening the crate, he found that there was an engine in it! There is reputed to have been quite a security panic in the company. A nice story but possibly more company legend than fact.

Which brings us to the arrival of our furniture in Niederbachem. It had been crated with as much care as if our few sticks of furniture had been an expensive Rolls-Royce aircraft engine. (I had told the company before I left that, compared with putting my family up in a hotel for several weeks and all the expenses of moving our furniture to Germany, it would have been much cheaper for us to sell the furniture we had and start afresh in Germany. Although that was indubitably true, the company's system did not permit of any such procedure.) At Niederbachem, we had already made the acquaintance of the Hausmanns family because Walter Hausmanns had laid out the garden at the house with the assistance of some of his sons. Walter came and helped us to take the crate to pieces so that we could get our furniture out and put it in the house. What to do with the wooden crate? Walter was delighted to have it and he used the wood to floor the loft in his house, which was quite close to ours. This was the beginning of a close friendship with the Hausmanns family, which has endured ever since.

Like us, the Hausmanns had four children although theirs were all boys. Walter, apart from being a very keen gardener in his spare time, was a colonel in the German army. He had been stationed in the Channel Islands during the war and was subsequently a prisoner-of-war near Newcastle-on-Tyne. His experiences had made him a great admirer of British democracy and, in addition to being a colonel in the army and a keen spare-time gardener, he was also very active in local politics and later became the local bürgermeister (mayor). Waldtraut, his wife, came from an aristocratic family and was actually a baroness, which did not prevent her from being an excellent housewife and cook. In general, the Hausmanns were just about as mad as the Stewarts - which is probably why we got on so well.

We were lucky enough to have been warned that there was a special custom when moving into a new home in that area of Germany. If you wanted to be very friendly with the neighbours, you invited them all in while the house was still in a mess and most of your goods and chattels were still in crates or boxes. You then dished out some simple refreshments in the form of beer and sandwiches, or the like. If, on the other hand, you preferred to lead a more reserved life, you did not invite them in at this stage. The neighbours would then respect the fact that you liked a quiet reserved life and relationships would be rather more formal.

Hugh, a friend of mine at the embassy had not been warned about this custom and was most annoyed when I told him. He was of a very gregarious nature and had formed the opinion that the Germans were a very standoffish lot - simply because he had sent the wrong signal. Most embassy staff were only in one location for a comparatively short period and it was up to the "permanent" embassy staff to let the "transient" diplomats know about local customs. Hugh felt that he had been badly let down.

For the Christmas break during our first year in Germany, we rented a holiday flat in Switzerland and set off with a heavily laden car. At the Swiss frontier, the Swiss policeman informed me - in German - that we could only import into Switzerland as much food as we could eat in one day. Without realising that Betty had not understood what he said, I said to her,

"We can easily eat the food we have in the car in a day, can't we dear?"

"What!" exclaimed Betty, "I have been slaving over a hot stove and we have a turkey, ham, Christmas puddings, Christmas cake and ..."

Perhaps the Swiss policeman did not understand English or perhaps he appreciated the joke. In any event, he simply waved us on.

This Swiss regulation - which has long since vanished, I hasten to add - was probably intended to ensure that visitors to the country had to spend money on food while there. It is the sort of thing that, rightly or wrongly, has given the Swiss a reputation among German people that is very similar to the reputation of the Scots among the English. The Germans have a nice story.

When the Lord God had finished creating the earth, the sky and the rest of it, He found that He had a few hours to spare and decided to make something really beautiful. He therefore created the Alps with the spectacular mountains, glaciers, lakes, Alpine pastures and so on. When He had done that, He admired His work for a moment and then thought it would be nice to have someone who could appreciate how beautiful it was. So He made the Swiss. There he was, a Swiss man complete with leather shorts, etc, etc. The Swiss looked around appreciatively and said "very nice". The Lord God said to him,

"Is there anything you want?"

"Yes", said the Swiss, "I would like a cow."

The Lord God promptly created a cow and the Swiss immediately set to work to milk it. When he had finished milking it, the Lord God enquired whether the milk was good to drink and if so, could He taste it? The Swiss obligingly handed Him some milk, which the Lord God drank with some relish. He then asked the Swiss,
"Is there anything else you want?"
"Yes", said the Swiss, "fifty centimes."
"What for?" asked the Lord God.
"For the milk, of course", said the Swiss.

When we first settled down in Germany, we tried sending the boys to German schools but rapidly came to the conclusion that the British and German education systems did not mix. As a result, the boys all ended up as boarders at the Royal School, Dungannon, in Northern Ireland. Although the separation was painful, it did have the good effect of tightening our connections with what I still tend to regard - even after all these years - as home.

Although the boys joined us in Germany for the main school holidays, this was not practical for the short mid-term breaks when they joined my parents in Bangor.

During one of these breaks, my father called Ian (No. 2 son) over to him and produced a school report.

"What do you think of that?" he enquired.

Ian looked through it with care and found all the usual comments such as "could try harder", "does the minimum necessary", etc.

"Well," said Ian, "I suppose it really is fair. Perhaps I should try a bit harder." My father looked at him very sternly and said,

"I don't think you have read that very carefully."

Ian read through it again but could still find nothing to which he could object. My father then said,

"How about looking at the top of the report?"

Ian did so and found that it was a report from the Bangor Grammar School on one Alexander Stewart and dated many, many years before.

Our daughter Siobhan was quite a bit younger and went to the local English school in Bad Godesberg run by a Mrs Day. Mrs Day was the leader of the "Ancient Britons" in the Bonn area. My friends at the embassy told me that most large cities abroad tend to have a group of "Ancient Britons". These are usually ladies who have outlived the native husbands they married, rigorously uphold English traditions and are frequently the backbone of the English church, if there is one.

I do not know what formal qualifications Mrs Day possessed but she was certainly an excellent teacher. Her little school, which ran in competition with another English school run by the embassy, had among its pupils many children from Commonwealth countries who had to learn English as a second language. She was particularly gifted in dealing with these children. One morning, a little African boy arrived at school in tears. Mrs Day comforted him and asked him what was the matter.

"I've just got a baby sister," said the little boy.
"That's no reason for crying."
"But she's black." said the little boy.
"Well, why not?" said Mrs Day.
"But I already have one who's black," said the little boy, "I thought when we were in Germany, I would get a white one."

Inevitably, my repertoire of funny stories was increased by a few German ones. One I particularly like was about the one-armed man who found shaving a bit tricky and one morning decided to go to the barber's shop for a shave. Unfortunately, the barber's shop had moved somewhat upmarket and had become a gentleman's hair stylist or something of the sort – a frequent occurrence at that time in Germany, as in England. Shaving was considered a bit infra dig in this refined establishment and the apprentice was allocated to do the job. He was not very skilful and the customer got chopped about quite a bit. The owner had seen what was going on and was somewhat worried. When the customer arrived at the counter to pay, he thought he had better chat him up.
"Have you been with us before, sir?" he asked.
"No," replied the customer, "I lost the arm on the Eastern Front."

Another story must have originated in the period between the two World Wars. It concerned a pub much frequented by ex-officers who sat around the "Stammtisch" and recounted their experiences of the First World War - deeds of incredible gallantry for which they were awarded various medals. At the end of the table was a very, very old man with a wrinkled face and droopy moustache who quietly drank his wine and smiled occasionally but without contributing to the conversation. Eventually, one of the younger men asked him politely when he had been in the army.

"Well," he said, "I was in the first Hanoverian cavalry from 1869 to 1872."

This awakened great interest; he must surely have been at the Battle of Sedan.

"Well," he said, "I was and I wasn't."

What did he mean, he was and he wasn't?

"Well," he said, "The first Hanoverian cavalry was in the reserve and we never came into action. We just stood around and brushed the horses."

Hadn't he killed anyone or anything?

"No," he said, "In fact, I never saw a Frenchman till it was all over and he was a prisoner."

How about medals and such like?

"Oh no," he said, "You don't get medals for brushing the horses."

A pregnant silence with the others looking rather embarrassed. Then, suddenly, he smiled. Looking round at them, he wiped away his droopy moustaches, took a sip of his wine and said,

"But we won. But we won..."

A word of explanation about the "Stammtisch". Most German pubs include a fairly large table at which only the very regular customers sit. Any interlopers daring to sit there are likely to receive - at the very least - a few dirty looks.

Betty and I were once touring in Switzerland and stayed for a few days at a hotel/inn in Lauterbrunnen. There was a tall, fair-haired waitress in the inn and I had commented to Betty that it was no wonder the Swiss were so prosperous - look at the way that girl worked! She was quite incredibly efficient and quick in everything she did. Because we always ordered in German and she replied in German, it was three days before we discovered that she came from Glasgow! From her we discovered that during that winter sports season, a fair proportion of the staff in the various hotels were young Brits, Americans, Canadians, Australians (or the like) who were having a look at the world before they settled down to pursue serious careers. A large number of them were due to finish in about a fortnight's time and they were plotting to commit the ultimate sacrilege by coming early and occupying the "Stammtisch" to the exclusion of the regulars. I should love to have seen it!

We attended the Anglican church in Bad Godesberg, where Betty and I became members of the choir. The church we used belonged to one of the German churches whose congregation had moved into larger and more modern premises. The organ was very old and although it was reputed to have substantial antique value, its musical capabilities were severely limited. The German owners of the church asked us to be particularly careful to ensure that the church was locked up promptly when we left in case vandals should damage the organ. The chap from the British embassy who played the organ when we first arrived was heard to enquire whether anyone knew any good vandals.

When he moved on, his place was taken by an American lady called Sue Felber. One Sunday morning, one of the foot pedals was sticking and as Sue's hands were fully occupied with the keyboard, I - as the nearest member of the choir - was given the task of freeing this foot pedal whenever it jammed. What the congregation thought I was up to when fiddling around between Sue's feet, I dread to think!

The Anglican church choir also formed the basis of a choral society called the "Rhine Singers" which gave quite a number of successful concerts in the area, sometimes in conjunction with German choirs. The fact that we tended to find difficulties with the same parts of the music gave an immense feeling of international solidarity.

This feeling of international solidarity was illustrated to me some years later when Betty and I were on holiday together in Venice. We attended a concert given in one of the churches there by a touring choir from Lübeck in North Germany. Part of their programme was a piece by Brahms in which I had, in fact, participated as part of a choir in England. The piece is unaccompanied and includes a particularly tricky bit where it is very easy for the choir to go flat. Sitting in front of me in the church was an Italian who obviously also knew the piece and as we approached the tricky bit, I could see him getting more and more tense (probably so was I!). When the German choir came to this awkward bit, they mastered it with consummate ease and the Italian in front of me relaxed and gestured with his hands to indicate his admiration - I knew exactly how he felt.

While on the subject of Italy, one of my younger colleagues at Rolls-Royce told me an amusing story about his bachelor days. A group of three of them were intending to go camping in Italy and they tried to recruit a fourth member of the party to fill the car and slightly reduce the expense. The chap they approached was not keen on the idea and complained that camping was a cold, wet, miserable business. They reassured him that camping in Italy was something quite different altogether and described how evenings camping in Italy could be spent sitting before the tent in the warm evening air drinking Italian wine.

When they crossed into Italy, they did so in a place where wine was still dealt with in the old-fashioned manner. Older readers may have noticed that Chianti bottles used to have a straw wrapping which included a straw handle. The wine was originally sealed with a small quantity of olive oil instead of with a cork. Provided the bottle was held carefully by the handle, the olive oil remained on top and, before serving, it was only necessary to throw the first drop away to get rid of the olive oil.

Unfortunately, our friends were not aware of this subtlety so, on their first evening in Italy, sitting before the tent, the wine was duly poured out. As it was dark, it was impossible to discern any subtle differences in colour and texture and, unfortunately, the new arrival in Italy got the olive oil.

"Well," he said with a grimace, "I suppose it's an acquired taste".

The house in Niederbachem had its drawbacks and, after about two years, we moved to another house on the other side of the river. This was in a very pleasant area known as the "Siebengebirge" or "Seven Hills". There was quite a steep slope on the road at the front of the house and there was a particular post on the other side of the road - part of a garden fence - which was invariably used by a certain dog when obeying the calls of nature. One morning, the weather conditions were such that the ground temperature was very much lower than the air temperature so that the drizzly rain froze as it touched the ground and eventually the whole area was covered by a smooth sheet of ice.

I happened to be at the kitchen window, which overlooked the road, when the dog approached its favourite post. With great difficulty, it struggled up the icy slope and managed to reach its favourite post but when it raised its leg and touched the post, this minimal extra force was sufficient to cause it to slide away from the post again for approximately one metre. It then repeated the procedure and I had time to call the whole family to the window so that we could enjoy the spectacle. The poor animal must have tried at least twelve times before it gave up and looked for a downhill post.

Living in Germany, the children had become acquainted with the German custom of the "Osterhase" or Easter hare. Early on Easter Sunday morning, parents hide Easter eggs in the garden of their houses for the children to find. These are supposed to have been laid by the "Osterhase" - a mythological mammal capable of laying eggs. Actually, I had never seen a hare in the area but one Easter Sunday morning, I was looking out of the same window and there was a hare gently loping up the road. It was going so slowly that I had time to call the children to look,

"There you are, there it is!"

My job in Germany involved substantial travelling to aircraft design offices in various parts of the country as well as to government establishments in Bonn and Koblenz.

On one occasion, there was some slight problem at the Dornier works on Lake Constance, which demanded the attention of a specialist designer from Derby. I accompanied him to the Dornier works where he soon sorted out the technical problems. In the evening, we were entertained to dinner by one of the Dornier engineers. It was high summer and he took us to a beautiful restaurant where we could dine in the open air with Lake Constance at our feet. It was a glorious evening with the sun setting over the lake and Switzerland visible on the other side of the lake. The chap from Derby was very impressed and commented that it must be really wonderful to live in a place like that.

"Oh I don't know," said the Dornier bloke, "In fact, there are two months between the end of the yachting season and the beginning of the skiing season when there is nothing to do."

Some people do have a hard life of it!

One of my colleagues had to do business with a German civil servant with an incredibly negative attitude. He never had any ideas himself and showed no enthusiasm for any bright ideas that anyone else might have. Discreet enquiries revealed that there was a very good reason for the civil servant's negative attitude.

During the war, while employed in some ministry or other, he had a bright idea. He had read somewhere about the bolas which is used by South American Gauchos for catching cattle and other creatures. The bolas consists of a rope with a weight at each end; the Gaucho swings this round his head and then releases it in the appropriate direction. It wraps itself round the creature's legs and brings it down. The civil servant's idea was to have a similar device consisting of a metal cable with a weight at each end. An appropriately trained pilot could launch this from quite a small aeroplane in the direction of overhead power lines. The cable would wrap itself round the power lines, short them out and interrupt the electricity supply over a large area

- particularly if strategically located power lines were chosen. He prepared a careful note describing his idea and submitted it to higher authority.

In due course, he was invited to meet some senior military officials in Berlin. They congratulated him on his idea but pointed out a difficulty. Whereas the majority of German power lines were overhead, a high proportion of British power lines were underground. Not only was it impossible to use the idea against what overhead power lines there were in England (because the British would then have knowledge of the idea and use it against the Germans) but they considered the idea so dangerous that they had decided that its inventor should be held incommunicado - so he was locked up for the rest of the war.

His lack of enthusiasm for bright ideas suddenly appeared more reasonable!

A story, which was going the rounds at the time, was about an RAF squadron that entertained a Luftwaffe squadron to dinner. It was a fairly formal affair with the ladies present. During the apéritif stage of the proceedings, the Brits mentioned that it was their custom for the youngest officer present to propose the toast to the ladies. A fairly rapid investigation showed that the youngest officer present was one of the Germans so he was allocated the job. He found an excuse to depart to the toilets for a few minutes, got out his German/English dictionary and sorted out – as he thought - a suitable translation for the equivalent German toast "Auf die Damen". In consequence, when the appropriate moment came, he rose to his feet and announced,

"Gentlemen! Upon the ladies."

One of the joys of living in the Bonn area, apart from the fact that that part of the Rhineland is itself very pretty, was the ease of access by car either up to the Dutch coast or down to the Alps and, with a little extra effort and time, down to Italy and the Mediterranean. It was also, of course, quite convenient to drive to France. Every two years, when the Salon Aeronautique took place at Le Bourget, Betty and I would drive over to Paris. I would usually take one or two days holiday so that we could drive across in a fairly leisurely manner to enjoy some of the countryside on the way. It was also, of course, useful to have the car available in Paris as extra transport during the show. Most of the representatives of the aircraft industry resident in the Bonn area did something similar at the time of the Paris air show.

This gave rise to a story about the Bonn representative of some English company who was visited by a friend from the head office in England. This chap was taking a few days leave so that he could travel by car with his wife through Germany and then through France to Paris in time for the air show. However, when he got to Bonn, there was a message waiting for him saying that there was a crisis and he must fly to Paris immediately to deal with some urgent business. What to do with wife and car? However, the Bonn representative was also going to the air show so he offered to deliver wife and car to Paris and then fly back to Bonn at the end of the show. They made all the necessary arrangements, checked through the green card and the other documents and had a letter of authority typed to permit the Bonn representative to use the car.

In due course, the latter arrived at the French frontier where the gendarme first examined the green card and the letter of authority. He seemed a little puzzled but the chap from Bonn said,

"C'est l'auto de mon ami."

The gendarme then examined their passports and looked quizzically at the different names.

"C'est la femme de mon ami," said our friend.

The gendarme smiled.

"C'est l'auto de votre ami. C'est la femme de votre ami. Monsieur a un ami merveilleux!"

We spent one summer holiday at a campsite on Lake Garda in Italy. This campsite was chosen on the advice of some American friends (the Felbers – I have already mentioned Sue) who were heading for the same place. John was a civilian associated with the American army and, in fact, the campsite was selected because it was next door to an "American Facility". This facility was supposed to be for the benefit of the American officers of an army stationed in the neighbourhood. The army had, in the meantime, been moved on somewhere else but the facility was still there and those who knew about it did not feel that it was any part of their duty to inform the authorities in Washington that it was no longer necessary. As guests of our American friends, we could also enter the facility, which had yachts, water skiing, tennis - you name it. We were also quite close to Verona and by making mutual baby-sitting arrangements (our tents were adjacent to one another) the parents and older children could visit the amphitheatre to see the shows there. Betty and I saw "The Sleeping Beauty" danced by the Kirov Ballet from Leningrad (as it then was) and the opera "Aïda". The first night we went, we paid for expensive seats down below but on the second occasion, we had the cheaper seats up at the top. Things may have changed in the last thirty years but we found the latter preferable. The customers up there were mainly local Italians who had brought food and bottles of wine with them to make for a really enjoyable evening and they were, I must say, extremely hospitable. The acoustics were a sensation. It seemed quite impossible that the soloists should be able to make themselves heard right at the top of this enormous amphitheatre, which holds about 20,000 people and is open to the heavens, but they can - clear as a bell.

During another holiday in Italy, we rented a villa at Lago Bracciano near Rome from friends at the embassy in Bonn. It was brand new and was supposed to have been finished - a gentleman called Alfredo being in charge of the building operations. When we got there, we found that there was no glass in the windows - which was not serious, it was scorching hot day and night - but, more importantly, there was no water and no electricity. We went to see Alfredo who launched into a long explanation in Italian finishing up with the word "domani[1]". Fortunately, my Italian was practically non-existent so my reply was simply

"No, Alfredo, acqua[2] oggi[3]".

Another long explanation in Italian finishing with "domani".

Reply, "Acqua, oggi"

Another long explanation in Italian finishing with "domani".

My reply remained, of necessity, very simple, "No, Alfredo, acqua oggi."

This went on for some time but, after a while, he came round to the villa and rigged us up with a water supply (rather Heath-Robinson but sufficient) from next door.

The next day, I was back to see Alfredo again.

"Alfredo, elettricità".

A similar rigmarole took place to that concerning the water on the previous day but my response was always the same (it was more or less all the Italian I knew),

"No, Alfredo, elettricità oggi".

After some time, he came round and rigged us up with an electricity supply - even more Heath-

---

[1] Tomorrow
[2] Water
[3] Today

Robinson but we were now equipped with water and electricity and settled down to enjoy our holiday. Lack of knowledge of a language can sometimes be an advantage!

I had to interrupt my holiday, however, because of a demand from the company that I attend to some urgent business in Munich so, leaving my family at Lago Bracciano, I flew to Munich. The job lasted most of a week but as some small compensation I succeeded in getting a ticket to the opera for one of the evening performances - it was the Wagner opera "Tristan und Isolda". The conductor was a man called Keilbert and, in fact, he had a heart attack during the performance. I did not actually see him fall because I was watching the stage but I heard the thump; the music then tailed off and the curtains were drawn. Herr Keilbert was carried off backstage and, shortly afterwards, a man appeared in front of the curtain to ask if there was a doctor in the house. Several people jumped up around the theatre and went to see what they could do to help. One of the people who jumped up was a lady doctor in my immediate vicinity. After a considerable delay, the lady doctor returned and said that although they had taken Herr Keilbert to hospital, she did not think he would survive. In due course, the official appeared in front of the curtain and told the audience that Herr Keilbert had been taken to hospital and that they were hoping for the best. Although there was a substitute conductor available, the cast did not feel that they could possibly continue singing that evening so the performance would have to be abandoned.

So we all trouped out of the opera house. Herr Keilbert's death was confirmed in the papers the following morning. When I got back to Italy, I related all this as a matter of great excitement to my family, who listened with appropriate interest. When I had finished, my youngest son - then aged about 10 - asked,

"Did you get your money back?"

In later life, he has been running his own business very successfully.

Lack of linguistic ability can be very rewarding at times and need not necessarily be genuine. On one occasion, I had brought my car (which was, of course, registered in Germany) over to London. I had parked it overnight at a point where parking was permitted until 9.00 am. When I emerged from the hotel, it was shortly after 9 o'clock and there were two policewomen taking a great interest in my car, carefully marking the tyres with a piece of chalk. I was not sure what the purpose of this manoeuvre was but suspected that it boded no good so I went up to the two ladies and said,

"Haf I somet'ing wrong done?"

They explained to me, very politely, the significance of the yellow lines and pointed out that I should not have been there after 9 o'clock so I said,

"What must I then do?"

They then informed me that there was another parking area round the corner where I could park without any difficulty but I replied,

"But now I go way".

Oh, in that case, they assured me, there was nothing to worry about and they hoped that I would have a very pleasant holiday in England.

One of the Rolls-Royce contacts in West Germany was a senior aircraft designer whom I got to know very well. While functioning in that capacity with one of the German aircraft companies, he was also working towards his doctorate. Needless to say, he was in regular contact with his professor. One day, the professor expressed an interest in the military aircraft for which he was the senior designer and suggested that he should write him a short note on the subject - not more than about ten pages - describing the main features and advantages. Needless to say, he obliged. The professor was well known in the industry and was in frequent contact with the appropriate ministry so there was no danger of any breach of confidence.

Shortly afterwards, the designer attended a meeting at the responsible German ministry and one of the officials there mentioned that they had obtained a consultancy report on his aircraft from a certain professor. Enquiry provided the name of the professor – our friend's professor. After some persuasion, he managed to obtain a copy of the professor's consultancy report from the ministry official.

Back at home, he punctiliously compared the professor's consultancy report with his original note. There was one comma extra!

He never did discover the amount of the consultancy fee.

Reverting to Munich for the moment, a rather amusing incident occurred there which I unfortunately missed - although I heard all about it afterwards. One of the Rolls-Royce directors was visiting Munich and was taken to what was possibly one of the most famous beer-drinking establishments in the world and associated, of course, with a famous brewery. Because he was an important visitor, his business hosts had arranged for the head brewer to come out from behind the scenes to meet him. Sitting down and having a beer with the visitors, the head brewer started to converse in remarkably fluent English with a strong American accent. He was asked how it came about that he spoke such remarkably fine English.

The brewer explained that, not long after the war, two American military policemen had turned up at the brewery and lugged him off to the American military headquarters where he was introduced to a gentleman who, it transpired, represented a large brewery in the United States. This gentleman offered him a job as chief brewer in this American brewery. Things were tough in Germany at that time and this was an offer that he certainly could not refuse. He received an excellent salary, a car for himself and a car for his wife, private school education for their children and very generous leave in Germany with all expenses paid twice a year. He then produced a full-page advertisement from an American newspaper - with a picture of himself, his wife and the children in Bavarian costume - announcing that Herr so-and-so from the famous brewery in Munich was now the chief brewer at this American brewery. The visitor enquired whether the Americans had really kept all their promises. In fact, they had even bettered them in some respects.

"If it was so good," enquired the Rolls-Royce director, "why did you come back?"

The brewer shook his head sadly.

"I couldn't stand the beer."

"But you were the chief brewer, why didn't you do something about it?"

"They wouldn't let me into the bloody brewery!"

This struck me as being a typically American stunt, which would not be found among Europeans but ...

Towards the end of the Sixties, I was in my office when the telephone rang. It was a Rolls-Royce director − not the one in the above story - who asked me if I was busy. Silly question. I was not terribly busy but that is not something to be admitted readily. When would I next be in Derby? I was due there the following Tuesday. That appeared satisfactory. Would I go and see him when I was in Derby? What is it all about, I enquired.

"We want you to go to Japan," he said and hung up.

I made some enquiries from other people in Derby and found out that they wanted me to go to Japan to help sell the RB 211 engine in the Lockheed Tristar to Japanese Airlines - in competition with the GE engine in the DC 10. This seemed to me to be quite crazy so I sent a note to the director who was my boss (not the director who had 'phoned me) pointing out that I did not speak Japanese, I knew nothing about Japan and that, furthermore, I did not really know very much about the RB 211 engine because I had been otherwise engaged during the critical period. My activities in Germany had been concerned with quite different engines. Despite all my efforts, I was given fourteen days to learn all about this engine before departing for Japan. During these fourteen days, I managed to derive one piece of sensible information. There was one very able engineer at Japanese Airlines who was making life difficult for the Rolls-Royce sales people by asking questions that they could not readily answer. For various reasons the Rolls-Royce Engineering Division declined to release any engineers to help so the sales people had looked around for an engineer with some appropriate qualifications who was under their control and not under the control of the Engineering Division. I answered the bill.

I was originally supposed to be going to Japan for quite a long period (six months to a year) so Betty did not want to stay in Germany on her own. Fortunately, our cottage in Breadsall had become vacant at just the right time, so that fitted in well enough.

Arriving in Japan, I soon met the Japanese Airlines engineer already mentioned. He was an extremely able engineer and a very pleasant man so we got on well together. After dealing with his questions for a week or so, I returned to the company's Tokyo office and told one of the permanent people there that it was quite obvious to me, from the questions I was being asked, that Japanese Airlines had not the slightest intention of buying either the Tristar or the DC 10. They were going to buy more jumbo jets with Pratt & Whitney engines because the questions I was being asked were simply such as to provide my Japanese opposite number with ammunition to throw at Pratt & Whitney.

"Oh yes," said the other Rolls-Royce chap, "We know that."

"If that is the case," I asked, "What on earth am I doing here?" (Language slightly modified for the benefit of younger readers!)

"It is quite simple," he said, "The Lockheed office in Tokyo also knows that Japanese Airlines are not going to buy any Tristars and are telling the headquarters on the West Coast that they are not going to get the business because Rolls-Royce is not trying. You, my dear Alex, are the evidence that Rolls-Royce is trying!"

Nevertheless, I quite enjoyed my time in Japan although, in fact, I was only there for about three months and not for the longer period originally expected. My efforts may not have been entirely wasted because I also took part in meetings with a Japanese internal airline who did eventually buy Rolls-Royce engines.

While in Tokyo, I met an American who was on a business trip in the Pacific area. He was very pleased with himself because he had managed to cross the International Date Line in such a way that he had two hotels bills with the same date from hotels more than a thousand miles apart. This, he was quite sure, would drive the Personnel and Administration Department at home quite round the bend!

There are, I believe, some people who find Japanese cuisine admirable. Personally, I did not and therefore appreciated the riddle that was circulating at the time.

"What's the difference between a successful and an unsuccessful diplomat?"

"A successful diplomat is one who, in the course of his career, has acquired an American income, a Chinese cook and a Japanese wife. An unsuccessful diplomat is one who, in the course of his career, has acquired a Chinese income, a Japanese cook and an American wife!"

Let me hasten to add that, from my rather limited experience, American women get a bad press - although not all American men would agree with me. I remember being in the Platzl, a Bavarian music hall in Munich, on one occasion with an American friend. The next table was occupied by a group of Americans aged somewhere around the fifties. There were two men (one of whom was the guide) and about eight women. I drew the attention of my American friend to what seemed to me an odd distribution of the sexes.

"Oh," he said, "they have worked Elmer to death, drawn the insurance and bought a fur coat and a trip to Europe."

Reverting to the subject of diplomats, I heard a rather nice story from one of my diplomat friends in Bonn. Before the war, there was a young British diplomat in Peking whose wife was extremely wealthy. In due course, this lady thought it would be nice to have her own private rickshaw and one day she spotted one which appealed to her on a second-hand rickshaw lot. It was beautifully decorated and had Chinese characters tastefully arranged on the back. She bought it and employed someone to pull her around in it.

In due course, the young diplomat was asked to go and see the ambassador who said that he was rather worried about the fact that the young diplomat's wife was being carted around in this rickshaw. The young chap explained that it was her own money and that there was really not a lot he could do about it. The ambassador, however, continued,

"I don't really object to the fact that the writing on the back of the rickshaw states that she is a licensed prostitute - but the price she is demanding really is ridiculous!"

At an earlier stage, I mentioned the Rolls-Royce Operatic Society. There was one chap who sang with us who had a particularly beautiful voice and was, in fact, a bus driver. In due course, he got a position with a leading operatic company in London and I heard from one of my friends that he had progressed from the chorus to singing small parts. However, after some time, I met him in Derby and enquired why he wasn't in London. He explained that he had given it up because he could make more money as a bus driver plus singing occasionally in working men's clubs.

While living in Germany, it became apparent to me that this state of affairs was not peculiar to England. It was during a performance of "Die Fledermaus" that this incident occurred. At the point in the operetta where the tenor (who has been locked up by mistake) was released, he handed a tip to the gaoler, who had treated him fairly well. The gaoler enquired where he worked and he said that he was a singer at the particular opera house. The gaoler then solemnly gave him his tip back.

Although most of my work in Germany concerned the aircraft industry, I was occasionally called upon to help with problems arising on the reciprocating engine side. There was a German fighting vehicle that was powered by a Rolls-Royce reciprocating engine and this was giving trouble. It is worth mentioning that Rolls-Royce only made the bare engine - everything else was the responsibility of other companies. As far as I could see, the vehicle had been designed by a committee and no-one appeared to have overall responsibility. However, be that as it may, some Rolls-Royce engineers came over from Shrewsbury and I went with them to help out with the investigations. We started off at a main overhaul depot where we carefully examined the equipment and the engine specialists became very suspicious of the cooling system (which was not made by Rolls-Royce). Another difficulty was that the weight of the vehicle had doubled since the original designer had picked the Rolls-Royce engine. However, the Rolls-Royce specialists thought that the most probable cause of the trouble was a leakage of oil into the cooling system. This would be deposited on critical surfaces within the engine, spoiling the heat transfer and damaging the engine.

We proceeded from the main overhaul centre to three German army units where these vehicles were in use. Our specialist said that we should visit the vehicle park, remove radiator caps and stick a finger in. When we pulled it out, we should then see whether the normal coolant could be quickly shaken off or whether a trace of oil was left. If there were a trace of oil, this would be clear evidence of what was causing the trouble.

At the first army unit, we went first to the workshop, where we met the German non-commissioned officer in charge. He appeared extremely efficient and his workshop was absolutely spotless. We then enquired politely whether we could go to the vehicle park.

"Of course," he said.

When we went round removing the radiator caps and putting our fingers in we did not find any oil but very frequently, to our surprise, there was no water there either so, at the next army unit, we went straight to the vehicles - with the same result.

We then set out for the third unit. In the meantime, the telephone had obviously been working and when we arrived at the unit, we found ourselves in a race to the vehicle park against a German private carrying two buckets of water.

Poor chap, with that handicap he didn't stand a chance!

(At this point, let me hasten to add that the appropriate German authority did not agree that the shortage of water in some radiators was the sole cause of the trouble.)

It was during this trip that I came across the story about the wine connoisseur who was asked to identify an unmarked red wine. After a quick sniff,
"It's Bordeaux, of course".
Another sniff - "Left Bank, fairly recent."
A small sip - "Fairly far upstream. H'm."
A slightly larger sip, then a broad smile,
"Good grief, how on earth did you get this? I thought the old devil drank it all himself. This is from old du Bois' place in Saint-Phillippe-les-deux-églises. His vineyard is on the left past the two churches. Well, well, well. He is quite incredibly behind the times. He doesn't even have a wine press. It's the two daughters, Marie and Clothilde, who do all the work."

A reminiscent look came over his face and he smiled,

"I have often seen them, Marie and Clothilde, treading out the grapes in the autumn."

Another rather more generous drink from the wine,

"This was Mlle Clothilde."

Another even more generous sip from the wine.

"Left foot."

It was also about this time that I heard the story about the man who wanted to be a waiter at

Claridges. This man worked as a waiter in some rather non-descript restaurant in London and, although he was reasonably well paid and the customers, the other staff and the boss were pleasant enough, he was ambitious and thought it would be wonderful to work somewhere really first class, like Claridges. So one day, he wrote to the headwaiter at Claridges asking what were the chances of getting a job. Rather to his surprise, he had a reply by return from the headwaiter at Claridges suggesting that he pop round some afternoon so that they could have a cup of tea together and discuss the matter.

He was warmly received by the headwaiter and, as they drank their tea, the headwaiter said,

"I was really very pleased to receive your letter and I'll tell you why. There are simply not enough Englishmen who take our profession seriously, which is the reason why the profession is largely manned by waiters from other countries. Having said that, however, I should inform you that the men who come to us as waiters have extensive experience and speak most if not all of the West European languages fluently."

Our friend looked rather downcast at this but the headwaiter continued,

"However, I did not ask you here just to discourage you. On the contrary, if you are really intent on making a first-class career as a waiter, I can write to my friend who is headwaiter at one of the hotels in Brussels and, on my recommendation, he will give you a job. Brussels is a good place to start because, apart from the large number of people of all nationalities who pass through - for obvious reasons - the country is itself bilingual, which produces its own opportunities and difficulties, which you should know all about. In particular, the local languages are French, which is of course very important, and Flemish, which is more or less the same language as Dutch, which is also very useful. If you work and study hard for about six months there, you should have acquired a reasonable knowledge of French and Flemish/Dutch, in addition to useful waiting experience. However, life is short and you should not stop there for more than about six months.

"Let me see, you will now have my letter of recommendation plus a good reference from the hotel in Brussels and with that, I should think, you will have no difficulty in getting into this hotel in The Hague in Holland." He started to make a list. "You will find that very useful, the Hague is a capital city with diplomats and so on passing through and the hotel is convenient to the railway station. In particular, in addition to perfecting your Dutch, you should make a determined effort to acquire a reasonable amount of German. You will find that many Germans pass through.

"At this point, you will have my letter plus two good references and will have substantially more freedom in choosing which hotel you would like to work in but I would suggest, in fact, that you now head up into Scandinavia and certainly go to Copenhagen first. I shall write you down the names of two or three suitable hotels and I am sure you will have no difficulty in getting into one or other of them. While there, you will of course learn to speak Danish and you will rapidly realise that the other Scandinavian languages - with the exception of Finnish, of course, which is a very different kettle of fish altogether - are very similar so not only will you be able to learn Danish but you will be able to master the other Scandinavian languages sufficiently in six months or so. Nevertheless, I think spending a bit more time in Scandinavia would be useful and I suggest you move from there up to Stockholm, where I can recommend a suitable hotel. By now, you will have my letter and references from three good hotels and I think you will have no difficulty in getting into this particular hotel. Stockholm is useful because quite a lot of Russians pass through there and, although it cannot be expected that you will master the Slavonic languages as well as the usual European languages spoken in Western Europe, a few words of Russian would certainly be very useful. Finnish, however, as I mentioned, is a different matter altogether. It is not even an Indo-European language but a few words, again, would be very useful.

"Let me see. By now, you will have my letter plus four good references and I would suggest that the next port of call should be Berlin. Again with the references you will have by then, you could take your pick but I will include a few suggestions in the list. While there you must, of course, perfect your German. Germany is now a very important country and fluent German is an essential.

"By now, you will have been working in a very concentrated manner for getting on for three years and although I should certainly not give you the impression that it would be a time to relax, it would be useful at this point to broaden out somewhat and for this purpose, I would suggest Vienna. Here again, with the references you will have, you can take your pick but by then you will know which one to go for. You may wonder why I have not mentioned Italy or Spain. The reason for that is quite simple, there are so many Italians and Spaniards working in the trade that you will have learnt Italian and Spanish as you go along.

"As I said, the purpose of the visit to Vienna is to broaden out. German you will already be able to speak fluently and the Austrian dialect is not very different so you will have time to acquire some cultural depth, which is absolutely necessary for a waiter in an establishment like ours. Ah, the Vienna Philharmonic, the Vienna State Opera ..."

For the moment, he looked almost philosophical but he rapidly pulled himself together.

"Six months should, however, suffice in Vienna and then, of course, you must go to France. With all your references and so on now, you can take your pick of whatever hotel you really want in Paris but whichever it is, you should stay there for at least a year. France is, in many respects, our Mecca.

"And after a year or so in Paris, you may care to write to me again."

So, after quite a large number of years, our friend was working as a waiter in one of the best establishments in Paris. He was very well thought of, was quite well paid, had a nice car and a beautiful French girlfriend. Both social life and career seemed very good but, nevertheless, he still had this idea that it would be good to work at Claridges so he wrote, once again, to the headwaiter at Claridges but this time including copies of all his magnificent references.

Once again, he received a reply by return in which the headwaiter stated that he was delighted to hear from his young friend again and that he had taken the most unusual step of putting him at the head of the waiting list.

A few weeks later, he received a letter from the headwaiter at Claridges saying that one of the waiters there had been knocked down by a bus and killed and they had an immediate vacancy, indeed an urgent vacancy, so if our friend could come immediately he could have the job. So our friend threw up his job in Paris, kissed his girl-friend goodbye, jumped in his car and headed for London.

On his first night on duty at Claridges, he was waiting in the kitchen for the first order. The headwaiter came in, gave him the order and said,

"Four ladies and four gentlemen at the table in a corner of the restaurant. Nouveau riche. Watch it!"

Our friend, while taking the hors d'oeuvres, noticed that they were indeed nouveau riche. Everything about them looked expensive but nothing was really in very good taste. One of the ladies, in particular, was wearing a dress with a quite incredible décolleté. She dropped her table napkin and instead of saying, "Waiter, fetch me another table napkin, please," she tried to pick it up herself. In the ensuing struggle, something emerged from the enormous décolleté which should not have emerged and the décolleté was so enormous anyway that she did not even notice!

Our friend, however, did notice and thought,

"If the other guests see this, it will give the place a bad name."

So, rapidly moving over to the table and with movements so fast that the eye could not possibly follow them, he picked up a soup spoon, whipped the offending item back into place and was gone.

At the kitchen door, the headwaiter was waiting for him.

"Oh no, Mr Smith, that will not do at all."

Our friend replied,

"I'm very sorry sir, but I did what I thought was best in the circumstances. But I am always ready to learn. What should I have done?"

The headwaiter drew himself up to his full height,

"At Claridges", he said, "one uses a warm spoon."

I told this story to some German business guests over coffee and brandy at the end of a dinner in a restaurant near Derby. When I had finished, the headwaiter came over and said that he could not help hearing the story I had told and he thought he should let me know that he himself had previously been a waiter at Claridges. He said that as far as he knew, the incident I reported had never occurred. Should such a thing happen, however, and he drew himself up to his full height:

"One would be expected to use a warm spoon."

The German guests were most impressed.

Looking back over this account of my period in Germany, it all seems remarkably light-hearted. In fact, the bright spots merely stand out against a background of some very hard grind – but the hard grind is not what this book is about.

The remaining part of my stint in Germany after I returned from Japan was not terribly happy and had probably best be skipped over quickly. In the first place, Betty had her first operation for cancer although, at this stage, it only involved the removal of a small part of her right breast.

The fusion between Rolls-Royce and Bristol Siddeley had meant that there was considerable duplication among overseas representatives and I did not feel that my position in Germany was particularly secure. As a result of this, we decided that it would be better if Betty remained in occupation of the cottage in Breadsall and, for my part, it seemed a good idea to brush up my engineering skills again. Although I had originally gone out to Germany as the company's technical representative, the job became less and less technical with the years and involved an increasing amount of liaison work. If I wanted to keep my hand in as an engineer, I would have to do something about it and so, to this end, I registered as an external research student at my old university and started doing some theoretical work on a slightly abstract subject which had interested me since my student days. This involved sacrificing a substantial part of my spare time to do the work - but it seemed wise to keep my hand in "at the trade".

On a February day in 1971, I was driving along the German autobahn heading for Dusseldorf airport, in order to fly to Birmingham for a business visit to Derby. I was listening to the German news bulletin and heard the announcer state that Rolls-Royce had called in the receiver. What to do? Well, I had an airline ticket to Birmingham so I might as well carry on and see if I could find out what was happening. The company had booked a hire car for me at Birmingham and, astonishingly enough, the representatives of the car hire company in Birmingham did not know (although their headquarters were in Derby) that Rolls-Royce had had to call in the receiver. They cheerfully handed over the car to me but when I checked it in at the headquarters of the hire company in Derby, the people there were anything but pleased.

When I went into the company the next morning, no-one had a clue what was happening. There was nothing useful I could do, I was owed a week's holiday from the previous year and as it was half-term, Betty and I pushed off with Siobhan, our daughter, to Gibraltar for a week. (Gibraltar was the only place in the sun which I could arrange to go to at such short notice.)

When I got back to Derby a week later, things had settled down a little bit and I was told to return to my post in Germany.

In the end, my return to Germany really only amounted to tidying things up before returning to Derby, where I was offered a job on my old trade of turbine blade cooling.

Although Austria was not part of my sphere of influence, I did have to go there occasionally for various business reasons and had become very friendly with Herr Markowitsch, our agent there. One of my last jobs before leaving Germany was to represent the company at his funeral, which was rather a sad event but - as is not uncommon at funerals - I did hear one rather nice story. At that time, the Chancellor of Austria was Dr Kreisky. One of his political associates was a particularly ugly man – in fact the Austrians said that if he was not the ugliest man in Austria, he must be the ugliest man in Europe. Need I say more? At any rate, the story was that Dr Kreisky and the other chap had died and were down in Purgatory. Dr Kreisky was getting the lot - whips, hot irons and goodness knows what else besides. However, about fifty metres away, he could see his associate sitting on a comfortable armchair with Brigitte Bardot on his lap, obviously having the time of his death. In due course, the foreman devil passed by and Kreisky queried matters, while making it very clear that he was quite sure that he deserved everything that was happening to him because of what he had done while on earth. On the other hand, he went on, as far as he knew the other chap had been no better than he was and - look what was happening to him and look what was happening to the other chap. The foreman devil smiled.

"You don't understand," he said, "That's not his purgatory, that's Brigitte Bardot's purgatory!"

Another Austrian story concerns the garden of a palace in Vienna. Among other things, this garden contains a beautiful female statue representing Venus, the goddess of love and a male statue representing the incredibly handsome Adonis. The two statues are about fifty yards apart and gaze at one another.

Mercury, the messenger of the gods would, on his journeys around the earth on behalf of the gods, stop occasionally in this garden to rest. He would observe these two statues gazing longingly at one another - in winter when it snowed on them, in spring, when it rained on them, in summer, when they became rather dusty, and in the autumn when the leaves blew around them. This was so year after year, decade after decade and, indeed, century after century.

One day, however, when Mercury was resting in the garden, he suddenly felt sorry for the two statues and spoke to them.

"Look you two", he said, "I am only a very junior god and my powers are rather limited. However, tonight between 1 am and 2 am, you can come down off your plinths and do what you have been dreaming about all these centuries."

About half past one in the morning, the caretaker was walking through the garden when he suddenly heard a noise in the bushes so he walked over in that direction and was just in time to hear Venus say,

"Alright Adonis, the half-hour is up. Now it's my turn. You hold the pigeon and I'll do to its head what it and its ancestors have been doing to mine for the last few centuries."

On my last evening in Germany, I went back for dinner to the hotel in Bad Godesberg where we had stayed when we first arrived in Germany. Apart from this initial stay, I had used the hotel restaurant extensively over the years for business entertaining and I had got to know the headwaiter fairly well. I explained to him that this would be my last evening there and perhaps, if he was not too busy later on, he could come and have a farewell drink with me. This he did and enquired where I would be living in England. I told him that I was returning to Derby. He was very interested and not only informed me that he used to be a regular supporter of Derby County but described the brilliance of Raich Carter, a much admired footballer just after the war. Apparently, he had been one of the prisoners of war who were kept in England for some time after the war finished and, in fact, he was doing some sort of useful job - on a farm, I think. This enabled him to earn some money and he used part of it to go to the football match every Saturday afternoon. He had also supported Derby County on their away matches. I expressed surprise that he could do that. It was, he informed me, particularly simple because he was a prisoner-of-war. He would simply don his prisoner-of-war uniform, get on the train to wherever the away match was taking place and, when the ticket collector arrived, he would indicate his prisoner-of-war uniform and say something like, "Officer tell me go to ..." and give the name of the town where the away match was. If the ticket collector argued, he appeared not to understand or speak any English. He always won, even if Derby County did not!

In fact, of course, he picked the right time to support Derby County, which did extremely well just after the war and even won the FA Cup. We must have been at the same matches on some occasions, sharing in our admiration for Raich Carter's brilliance. Our little chat was a pleasant close to my stay in Germany.

Back to Derby

Before we reoccupied the cottage in Breadsall, we had had substantial changes made to it and, from being a modest agricultural labourer's cottage, it had became a reasonably substantial house. As we bought it, the cottage was really old, probably seventeenth century, and seems to have been part of what was once a much larger establishment.

At the turn of the year, after I had re-established myself in Derby, two of my close German associates died very suddenly. Their funerals were, as it happened, on the same day - one in Bremen in north Germany and the other in Lindau in south Germany. I went to the funeral of the latter on behalf of the company and, in a way, that represented the end of an epoch for me.

The funeral I went to was for an associate of the company who lived in Lindau, on Lake Constance near the Austrian frontier. His name was Helmut Sachse. It is, I must say, a matter of great regret that this man never wrote his memoirs because some of the incidents he revealed of the period before and during the Second World War were quite fascinating. However, the humorous event occurred much later when, together with another Rolls-Royce man, he and I visited an aircraft firm in Hamburg.

The convenient way to reach this firm was to take a taxi from the centre of the city to a point on the Elbe and, from there, to take a ferry across the river.

The meeting went quite successfully and, at the end of it, the senior representative of the aircraft firm suggested that instead of returning the way we had come, we should take the steamer that travelled up the Elbe to the centre of the city passing numerous shipyards and other places of interest on the way. He consulted his secretary, who went away to make some enquiries and came back, giving him a piece of paper. He read the result - in German of course.

At the steamer station on the riverbank by the aircraft factory, we discussed what he had said. I interpreted what he had said as being that the steamer went every hour at quarter past or twenty minutes past the hour, depending on the state of the tides.

Helmut told me that I had misunderstood the German and that what he had said was that the steamer ran every hour on the hour and stopped at 15.00 hours or 20.00 hours, depending on the state of the tides. I told Helmut that although I had not the slightest doubt that his German was much better than mine, I had been brought up by the sea and I could not imagine any timetable which could be affected by the tides in the manner he suggested.

Three o'clock - no steamer. Ten past three - no steamer. Helmut strode off along the quay to the office and came back furiously angry.

"They must have altered the timetable," he said.

I am afraid the other Rolls-Royce chap and I laughed. The ship turned up slightly later.

To be fair to Helmut, he often told this story to his colleagues as a joke against himself.

Because Helmut died over the Christmas break, the news of his death came to us rather late - in fact I received a Christmas card from him when I already knew that he was dead. The only way I could possibly get to Lindau in time for the funeral was to fly to Zurich and take the train to St Margrethen on the Austrian border. I then took a taxi through the little bit of Austria that touches Lake Constance in order to get to Lindau.

The train from Zurich was the last one of the day and, instead of a dining car, there was only a snack car. However, I did get a reasonable meal, which concluded with a very nice piece of Swiss Gruyere cheese.

Sitting beside me was a Swiss gentleman who politely enquired if I was enjoying the cheese. In order to pass the time, I said that it was very nice and that if the Swiss exported some of their best cheese instead of eating all the best stuff themselves, they might do more business in Great Britain - anything for a bit of fun and to make conversation. He carried on the conversation in a very pleasant manner, continuing to discuss the merits of cheese until he got up at St Gallen to get off the train. Then, before he left, he handed me his card with a smile and disappeared. I looked at the card:

Herr ...
Chairman
Swiss Cheese Export Corporation.

The joke was on me - I am sure Helmut would have appreciated it.

Back in Derby, I was soon in charge of a department of some 20 to 30 engineers and life was very different from that in Germany, where I had operated very much on my own. Certain things happened, furthermore, which made it clear that the years were going by. One day, I was discussing an engineering problem with one of the section leaders in the department. He had proposed a certain solution and I demurred because, as I pointed out, I had tried something very like it before and it had not worked. When he enquired where this had occurred, I told him that it was while I was working at Ruston and Hornsby in 1949 or soon afterwards.

"That," he said, "was a good year."

When I enquired why, he told me it was the year in which he was born!

Another straw in the wind was a management course organised by the company and taken by an American. One of the course documents, which we were supposed to read beforehand, was called "The middle-aged manager". Something for me, I thought but when I read it, I discovered that Americans regarded 35 as middle-aged. At over 45, I already had one foot in the grave!

The spare-time work I had undertaken while in Germany, as an independent research student of Belfast University, certainly paid dividends because it had freshened up my mathematics and such like and put me in a much better position on my return to the technical field. Combined with a very demanding responsibility in a field where the company had its troubles, pressing on with this external research project made heavy demands on my spare time and a considerable consumption of midnight oil was necessary.

Running a department did, of course, involve inevitable brushes with the bureaucracy. The offices inhabited by my department were on the third floor of a modern air-conditioned building in which it was impossible to open any windows. I was approached one day by the trade union representative who said that he and the chaps were worried that in the event of a fire, they might not be able to get out. He also pointed out that there had never been a fire drill in the building. I accordingly passed on his worries in a memo to the Personnel and Administration Department and in due course received a reply which stated, more or less, that the problem had been given due consideration by those best qualified to judge and that it had been decided that no fire drill was necessary. I wrote back expressing my thanks for the note but stating that as a precaution, I would keep my original note and his reply at home so that, should there ever be a fire, they would remain available as evidence.

Shortly after that, I went on holiday and, when I came back, the union representative approached me with a grin and informed me that there had been a fire drill while I was away.

In the meantime, the children were growing up and had become young adults. Meal times had always been important occasions in the Stewart household and frequently lasted quite a long time while we discussed all sorts of matters. The Encyclopaedia Britannica was always kept in the dining room and, if there seemed to be a lack of hard information on some particular topic, I would delegate the task of consulting the encyclopaedia to one or other member of the family.

One day, we were eating a meal consisting largely of salad and I was giving my opinions on some subject or other (pontificating is how the children described it) when suddenly the youngest son, Andrew, interrupted me,

"Dad!"

I reproved him gently and pointed out that I was speaking just at that moment but he would have his turn presently. I then continued, eating a little salad from time to time. I eventually informed Andrew that it was now his turn.

"It's too late now," he said, "You've eaten it!"

My private studies eventually permitted me to produce the necessary thesis and I was awarded a Ph.D. This happy event was, unfortunately, clouded by the fact that Betty's troubles reappeared and she had to have the rest of her right breast removed. She was not fit to travel to Ireland but insisted that I go over for the ceremony rather than graduate "in absentio". Unfortunately, my mother was also in hospital at the time so that put rather a damper on things as well. However, my father and three cousins were able to come. It was an unusually beautiful day in Belfast although, among all the young graduates, I did feel rather like the ghost at a wake.

The Seventies did represent a period of useful endeavour but, in retrospect, it seems to have been mainly a matter of grim determination, best skipped over rapidly.

Towards the end of that period, the main technical problem with which I had been concerned had largely been solved. I was then into my fifties and not looking forward to a final ten years at Rolls-Royce with no reasonable possibility of further promotion and with younger chaps, who were quite capable of doing my job, coming up behind me. By that time, furthermore, the mortgage on the house was paid and the children were more or less independent. Time to make a move.

Before leaving Rolls-Royce, however, it may be worthwhile to recount one amusing incident.

At a technical meeting, one of those present put forward a proposition that was quite obviously wrong. Instead of admitting his mistake, he tried to argue himself out of the position and the more he talked, the deeper he got into a mess – to the amusement of most of the others present. Eventually, the chairman - who was a Yorkshireman but usually spoke a rather correct English - interrupted him in a broad Yorkshire accent.

"Look lad," he said, "As we say in Yorkshire, 'when tha' gets tha' feet in t'muck, don't paddle abaht'".

Self-employed

On leaving Rolls-Royce, the idea was to do consultancy work based on the high technology I had learnt at Rolls-Royce, backing this up - if necessary - with technical translation work. In fact, I soon found that there was an almost unbridgeable gap between the high-technology part of industry and the general level of engineering. As an example, one day my telephone rang and the chap at the other end asked for Dr Stewart. He had somehow heard of me and had been told that I should be able to help them with a particular problem.

(At this point, it should be emphasised that the relationship between a consultant and his client is confidential. For this reason, the following story is distorted to ensure that the identity of the company and the nature of the problem are concealed - while the humour is, I hope, retained.)

What was the problem? It appeared that it was frightfully confidential but after probing a bit, I found that it was something to do with the goods they manufactured. I agreed to see the company a couple of days later. When I arrived, they showed me round the factory where they were very successfully manufacturing the goods in question. What was the problem? Perhaps we ought to go up to the office. There I was introduced to another of the directors and to George, who did the drawings. After further great emphasis on the confidential nature of the enquiry, it was eventually revealed that they were considering manufacturing a product that included an elliptical component. It was not exactly an original idea but seemed quite feasible.

What was the problem? There was another pause and some talk of confidentiality and then eventually George blurted out,

"How do you draw an ellipse?" The scales were beginning to drop from my eyes.

Not all the consultancy jobs on which I was involved were as silly as that one. In at least one case, I saved a company a very substantial sum of money but I was never made very welcome by the engineers employed by the company simply because they regarded my very presence as a criticism of their own efforts. (This is some time ago and I believe things have now changed). Another difficulty was that consultancy jobs tended to take quite a long time and were followed by a considerable delay before the bill was eventually paid.

In consequence of this, I concentrated more and more on translation. Although there are plenty of people with linguistic qualifications who can translate literary work, technical people with the necessary grasp of foreign languages are comparatively rare. In consequence, I found that there were quite adequate amounts of technical translation work to be had at reasonable rates of pay and, possibly more importantly, the individual jobs did not take very long - and the customers paid rapidly.

Working on my own from home had the great advantage that I could spend a lot more time with Betty and that she could help me with the administrative side of the business. In addition, she could travel with me whenever visits were necessary to customers at home and in Germany. In view of the doubt as to when the cancer might strike again, this was a great advantage.

At this time, Betty's mother was a widow and my father was a widower and we took this into account in planning some of our holidays. Not that we indulged in foursome holidays, the tastes were too disparate for that. On one occasion, Betty went with her mother to the Holy Land while I was able to satisfy one of my father's ambitions by taking him on a Rhine cruise - a division between saints and sinners, so to speak.

The Rhine cruise was on a British ship with British customers and a British crew (with the exception of the pilot, who was Dutch) so those on board did not have to worry about language problems. One of the ladies on board had some trouble with her camera, however, and wished to take it to a camera shop at the next port of call so that it could be opened in the darkroom to check what the problem was. Knowing that I could speak German, she approached me for help so I wrote her out an appropriate note to give to the man in the camera shop. She looked suspiciously at the German I had written and wondered what I had really written but I consoled her by telling her that I had said that the lady had a problem and would like him to fix it for her in the darkroom.

When she got back, I asked her how she got on. She sighed and said he only fixed the camera.

The Rhine cruise started and finished at Cologne so we combined it with a visit to our old friends, the Hausmanns family in Niederbachem.

Rather than drive back through Holland, which was the way we had come, we decided to drive to Trier and then over the frontier into France, where my father had never been.

The first day's journey took rather a long time before we found a suitable hotel in France and after a quick wash, we went down to the dining room. I asked my father if he would like a drink and he said that he was dying for a cup of tea. In consequence, when the manageress came, I said

"Du thé, s'il vous plait, mais dans la manière anglaise - avec du lait frais."

"Mais naturellement, Monsieur."

In due course, the tea arrived, complete with some fresh milk. I poured a cup of tea for my father and he sipped it with delight. Then, in his strong North of Ireland accent, he said,

"You know, Alex, I've always heard that the French were great cooks. It's absolutely true, this is a wonderful cup of tea."

He did, however, have occasion to really appreciate French cuisine the next day. We were getting fairly close to Boulogne (where we were to catch the steamer) but we had time to stop for lunch. I had a fairly modest sum of French money on me and did not really wish to change any more money because once we had had lunch, there were no more expenses to be paid in francs before we boarded the ship. We went into the restaurant and I spoke to the manager, showed him what French money I had and asked him if he could feed us for that. The fact that I had an "old gentleman" with me seemed to touch his kind heart and he agreed immediately. We ate like kings and had a very good bottle of wine as well.

I include this episode for the benefit of my children so that they will be aware of the advantage associated with taking an elderly relative to lunch.

My father had two other great ambitions. He wanted to go to Zimbabwe to see my brother and to meet the daughter-in-law, the two granddaughters and the great-grandchild whom he had never seen. He also wanted to do a tour through the highlands of Scotland and the Western Isles.

We managed to help him to fulfil both these ambitions - a small recompense for all the sacrifices he had made to help me when I was young.

While we were in Zimbabwe, my niece (who studied law in Cape Town) told us an interesting, and allegedly true, story about a South African judge. This story appealed to me because, as a professional translator, I was only too well aware that any foreign language - no matter how well you think you know it - is full of traps. At any rate, this judge's mother tongue was Afrikaans but the accused on this occasion was English-speaking so, before passing sentence - and waving a severe judicial finger at the accused,

"You," he said in his best English, "are the sort of man who likes to have a finger in every tart".

Like most non-physicians with a doctor's degree, I try to keep my title well clear of the medical profession. When my father was taken seriously ill, I caught the first available aircraft to Northern Ireland. After I had left, the hospital rang and my secretary answered the 'phone. Referring to me as "Doctor Stewart" (the title was always used for business purposes - it is worth at least 10% on the invoices), she said I was on my way. When I got to the hospital, I was met by a young doctor who introduced himself by his first name and then enquired if I had been at Queen's (as the University in Belfast is usually called by the locals). When I replied in the affirmative, he asked me if I had known a particular professor. The penny should have dropped at this point but, unfortunately, one of my sons had been at boarding school with a son of this professor and they had been bosom pals, so I responded with this bit of information. He then provided a variety of clinical details and generally dazzled me with science.

I visited the hospital every day and I was continually amazed by how well I was received by the hospital staff. Nothing was too much trouble. Eventually, one of the nurses said to me,

"You're the GP, aren't you?"

"No," I said, "I'm not a GP."

"But we were told that you were Dr. Stewart."

"I am Dr. Stewart but I am an engineer."

"Ah, that explains it. We all thought you were a bit thick!"

My father died after a mercifully short illness in 1985. We were under the impression that he was 90 because he had celebrated his 90th birthday the year before. However, when we registered the death and looked at his birth certificate, we found he was only 89. He had taken the precaution of celebrating in good time. He was buried with my mother in the new cemetery on the outskirts of Bangor, from which there is a fine view out to sea. That seemed very appropriate. "Crossing the bar" was one of his favourite poems.

> "Sunset and evening star
> And one clear call for me!
> And may there be no moaning of
> the bar,
> When I put out to sea,
>
> But such a tide as moving seems
> asleep,
> Too full for sound and foam,
> When that which drew from out the
> boundless deep
> Turns again home.
>
> Twilight and evening bell,
> And after that the dark!
> And may there be no sadness of
> farewell,
> When I embark;
>
> For though from out our bourne of
> Time and Place
> The flood may bear me far,
> I hope to see my Pilot face to face
> When I have crossed the bar."

One holiday Betty and I had was in Glengarrif in West Cork, a place with happy memories from the time when we lived in the Irish Republic. We took the opportunity of driving over to Kenmare where my old friend Father Long, now Archdeacon Long, was parish priest. After dinner, he invited us back to the parish house where we chatted in his study. On the wall, there was a large photograph of himself and the Pope. The archdeacon had been paying a visit to Rome a couple of years before and had called at the Irish college where he had been a student. As it happened, the Pope visited the college while he was there and he was introduced as being the oldest old boy of the college. Someone had taken the opportunity of photographing them.

"I find it very useful," he said, "Like every other parish in Ireland, I have great difficulty in raising money for the Peter's Pence[*]. Whenever one of the parishioners asks me 'What was the Holy Father saying to you, Archdeacon?', I tell him he was asking me which of my parishioners contributes to the Peter's Pence!"

Ever since my return from Germany, I have been a member of the local church choir and various choral societies.

One of the choirs included a bass singer who had reached the age of 86 but still had a magnificent bass voice. He had an illness, which

---

[*] A contribution made by Roman Catholic parishes to help defray Vatican expenses

required some time in hospital, and then he convalesced in a resort on the south coast. He sent the choir a card stating that everything had gone well and that, after his convalescence, he felt like a 21-year old. However, he went on,
"I can't find one that will have me."

Our children having grown up, as already mentioned, grandchildren began to appear in due course. One day, I was showing my grandson round the village church and he showed great interest in the electronic organ. (He was interested in anything electronic and is now a computer specialist!) The organ was being played at the time by my neighbour Willis Carter. He had been the church organist much earlier but had then retired. However, we were without an organist at that time and he had been persuaded to return and do the necessary until we could find someone else. By then, he was eighty-three. Willis was pleased that Mark, my grandson, was showing so much interest in the organ and demonstrated all the bits and pieces.
"Do you play the piano or anything like that, yourself?" he asked.
On receiving a reply in the negative, he said,
"Well, my advice to you is to take it up and, if you practise hard, by the time you're eighty-three they'll find they can't do without you!"
Willis played regularly for services at the crematorium. During his last illness, his relatives were sitting round the bed in the hospital. Willis was fairly heavily sedated and must have had a

powerful dream or hallucination. At any rate, he suddenly sat bolt upright in the bed and said,

"It's no good. It's time I was at the crematorium."

The family, realising that he must be dreaming about playing the organ for a cremation service, were rather amused, much to the horror of the patients round about who did not, of course, know the background. He died soon afterwards and was much missed in the village.

A few years later, Betty had to have another operation but there was little the surgeon could do. The cancer had gone too far. In the time that was left to us, we did what we could. Betty had always wanted to go to Corsica because her family tradition was that they had originated from there. The weather was kind and we stayed in a small hotel where one of the guests was an accountant. He had a nice story concerning his profession.

The managing director of a company was worried about the capabilities of his fellow directors and set them, as a small test, the problem of what two times two produced. The production director said that the answer would be 4 +/- .005 with a scrap rate of 5%. The technical director did a complicated calculation on his pocket calculator and came up with a result of 3.9999997 depending on the day temperature and the coefficient of linear expansion. The accountant said, "What would you like it to be?"

While on the subject of accountants, it may be worthwhile to include a story that comes from

No. 3 son, Andrew, who runs a business as a systems analyst in British Columbia. The story concerns a systems analyst and an accountant (people who traditionally do not get on) who were on a hiking holiday together in a forest area in British Columbia. They had been followed for several hours by an extremely large, hungry-looking and nasty-looking grizzly bear. Eventually, the systems analyst stopped, took off his rucksack, opened it and took out a pair of running shoes. The accountant said,

"You don't imagine that you can run faster than that grizzly, do you?"

The systems analyst replied,

"I don't need to run faster than the grizzly, I only need to run faster than you."

The last holiday Betty and I could take together was a tour by car through France and Germany. We managed to pull in visits to La Rochelle and the Lascaux caves in the Dordogne as well as visits to friends in the Massif Centrale and to the Guidets in Paris. We then drove eastwards into Germany to attend the wedding of the Hausmanns' eldest son Ulf before returning to England.

The cancer eventually caught up with Betty at the end of the decade. She was a great fighter and, right to the end, never lost her sense of humour. She died the day after our wedding anniversary (which was not long before Christmas) and when, a few days before, I asked her what she would like for an anniversary present, she replied - knowing well that the end was at hand,

"I want you to buy a small and very highly coloured watering can which can sit on the kitchen window sill. Perhaps you will then notice it and remember to water my indoor plants."

When the end was very, very close, we were talking. She was quite resigned to the fact that the end was at hand but said that she had only one regret and that was that she would not be able to attend the midnight service in the village church on Christmas Eve. I tried to comfort her by saying that she would hear the service sung by a much better choir with no one singing flat and nobody scooping. She smiled.

"I don't know, I think I will just tune into Breadsall."

"Lo some we loved, the fairest and the best
That Time and Fate of all their Vintage prest
Have drunk their Cup a Round or two before
And one by one crept silently to Rest."

Epilogue

The death of my wife Betty seems a good place to conclude matters for the present but one subsequent item may be worth a mention.

A few years ago, I decided that it was time I paid a visit to Bangor. I had not been back for some years and I knew that there had been great changes in the town and that the old sea front had been completed remodelled to produce a rather luxurious marina.

I stayed at my cousin's house and, on the first morning, I set off on foot to have a good look round the town. There were indeed many changes and much to see but, in the midst of my peregrination, the skies opened in typically Irish fashion. Taking shelter in the nearest pub, I ordered a beer and waited for the rain to subside. The time went by and still it poured and I started to get worried because I was due back at my cousin's house for lunch at 1 o'clock. About half past twelve, I went up to the bar and asked the girl to order a taxi for me. She lifted the phone, did the necessary and then told me, "He'll be round in about fifteen minutes, sir."

About fifteen minutes later, there was no sign of the taxi so I went back to the bar. She rang again and informed me,

"He'll be here in a wee minute, sir."

About ten minutes later, the taxi arrived and as I climbed in, I said with - as I imagined - heavy sarcasm,

"That was a very peculiar fifteen minutes."

The taxi driver smiled,

"It was indeed, Sir, I haven't seen it rain like that in a long time."

"Alex" I thought, "You're home. You can't win."

On another day, I walked up from the sea front through the centre of the town and on towards the old Abbey Church. The demesne wall has gone, Bangor Castle now holds some local authority offices and a substantial part of the erstwhile demesne is now a public park. The Abbey Church, however, is still there so I went into the graveyard and eventually found my great-grandfather's grave. It is marked with a little stone and there is nothing on it except

A STEWART

I told my children that when I die, they should waste as little money as possible but get me cremated in a cardboard box and then, when someone is going over to Bangor for other reasons, they can just insert the ashes in front of this little gravestone. The name is correct and that should save a lot of money. I do not, for one moment, imagine that they will take the slightest notice and anyway, I am still looking forward to some more funny things happening on the way to the graveyard.

While on the subject of final disposal, I have carried a donor card for many years but it is no longer much use because my organs are – at least for donor purposes – probably past their use-by date. I have been considering leaving my body to the medical profession, should it be of any use to them. Discussing this with some members of my family, my granddaughter – who is studying forensic science at university – supported the idea strongly. She said she would just love to practice on my bits and pieces!

ISBN 1412029740